Andre Löffler

Constrained Graph Layouts:
Vertices on the Outer Face and on the Integer Grid

Andre Löffler

Constrained Graph Layouts

Vertices on the Outer Face and on the Integer Grid

Würzburg
University Press

Dissertation, Julius-Maximilians-Universität Würzburg
Fakultät für Mathematik und Informatik, 2020
Gutachter: Prof. Dr. Steven Chaplick, Prof. Dr. Alexander Wolff, Prof. Dr. Sabine Storandt

Impressum

Julius-Maximilians-Universität Würzburg
Würzburg University Press
Universitätsbibliothek Würzburg
Am Hubland
D-97074 Würzburg
www.wup.uni-wuerzburg.de

© 2021 Würzburg University Press
Print on Demand

Coverdesign: Andre Löffler

ISBN 978-3-95826-146-4 (print)
ISBN 978-3-95826-147-1 (online)
DOI 10.25972/WUP-978-3-95826-147-1
URN urn:nbn:de:bvb:20-opus-215746

Preface

This book studies algorithmic problems from the *graph drawing* subfield of computer science. At a high level the field concerns being given an abstract graph and realizing it graphically by placing the vertices at explicit coordinates and providing curves realizing the edges. The coordinate space is primarily 2-dimensional (as to display the graph on a screen), and secondarily 3-dimensional. To make such realizations meaningful, restrictions are used, e.g., on the topology of edge crossings or on coordinate precision.

The contributions herein involve several mathematical perspectives: graph-theoretic, algorithmic, heuristic, and geometric. The problems studied here are split into two parts (Chapter 1 and Chapter 2 introduce and set-up the context).

The first part contains two chapters (3 and 4) on drawings of graphs in the plane where the vertices are required to occur on the boundary of a simple polygon and the edges are drawn inside the polygon. Chapter 3 provides graph-theoretic results on natural drawing styles and these structural results are leveraged to provide algorithms (via a logic-based algorithmic meta-theorem: *Courcelle's Theorem*). Chapter 4 highlights a new perspective (drawing edges with few bends) on the well-studied problem of *partial planar drawing extension* and provides natural first steps via efficient algorithms.

The second part concerns two problems (one spanning Chapter 5 and Chapter 6, and the other in Chapter 7). The common themes are that the given graph is *embedded* (a cyclic ordering of the edges around each vertex is given and the output drawing must respect this), vertices occur at integer coordinates, and edges are realized as line segments. Chapter 5 and Chapter 6 concern a natural aspect of graph drawing: *snapping/rounding* an arbitrary precision drawing to integer coordinates. Computational hardness, an exact approach (via integer linear programming), and a new practical heuristic approach to natural instances of the problem are given. Chapter 7 is, in my opinion, the strongest result of the thesis. It concerns an analogous result on *orthogonal polyhedra* to *Cauchy's Rigidity Theorem for convex polyhedra*. The proof of the rigidity result is constructive (via a simple, but novel, combinatorial 3-coloring algorithm) and involves rather detailed geometric analysis and imagination—it is the fruit of two years of discussions.

Having worked directly with Andre on several results herein (and supervised him throughout), it has been an enjoyable ride. His creativity and humor kept our meetings fun and productive—though the puns were painful at times. For a retelling of some of that ride, I invite you to read this thesis and I hope that you enjoy the results, and look to the conclusions for some interesting questions for further study.

Steven Chaplick,
Dept. of Data Science and Knowledge Engineering,
Maastricht University, the Netherlands.

Contents

Chapter 1

Introduction

"Within a graphic standard, a graph has infinitely many different drawings. However, in almost all data presentation applications, the usefulness of a drawing of a graph depends on its readability, i.e., the capability of conveying the meaning of the diagram quickly and clearly."

– taken from Guiseppe Di Battista, Peter Eades,
Roberto Tamassia & Ioannis Tollis, 1994 [DBETT94]

The task of producing high-quality visualizations of information involves a spectrum of challenges due to it being a craft as well as being an art form. Historically, visualizing data was manual work done by scholars and experts, and the material used was very expensive. Thus, significant effort was put into creating *pieces of art* – a typical example can be found in Figure 1.1 (a): This geographic map of the region around Würzburg was enriched with depictions of prominent people and places of interest to illustrate the data at hand. The main body of water is shown as an oversimplified oval shape, with islands on the inside and all ports placed around it, preserving their cyclic order while ignoring real-world distances and geographic accuracy. Today, *technical schemata* such as UML diagrams are regularly used in software engineering and project management to convey complex information quickly and without ambiguity: The nodes contain information about the entities they represent, whereas the links between the nodes are annotated with the type of connection. In some cases, designers intentionally blur the lines between craft and art, creating beautiful schematic representations such as the metro map shown in Figure 1.1 (b).

Nowadays, there is a plethora of information to visualize. It ranges from the abstract data of social networks to blueprints and flowcharts to geographic data. Each domain is asking for its own meaningful and pleasant visualization to assist the viewer in capturing all relevant information. Sometimes, tremendous effort is spent drawing large networks by hand: See for example the Human Metabolism Map shown in Figure 1.2 – the layout was crafted by five people over the course of more than a year. This drawing has proven to be an effective instrument to a team of researchers around Thiele [TSF+13]. Unfortunately, not every drawing of such size can be hand-made by experts.

In particular, the need for *good* automated drawing algorithms is clear and is a primary focus of the, now roughly thirty year old, field of *graph drawing*. Graph drawing focuses on discovering the structural properties of different classes of networks, exploiting them to a develop graphic standard tailored to each class respectively. In graph drawing,

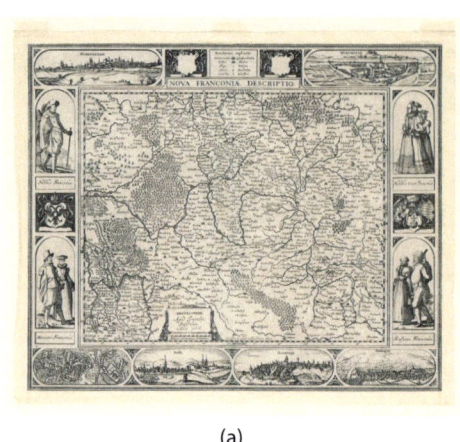

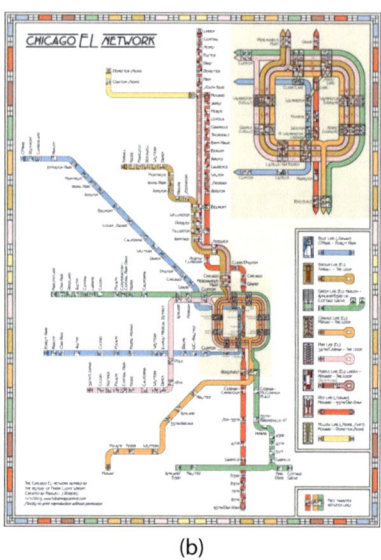

Figure 1.1: (a) Historic map "Nova Franconiae descriptio" of the region around Würzburg, created 1626, produced by etching (Licensed under CC BY-NC-ND 4.0 by the Würzburg University Library; found at http://vb.uni-wuerzburg.de/ub/permalink/36gfm912139_105787462, Signature 36/G.f.m.9,12,139). (b) A map of the Chicago public transport network, created by Maxwell Roberts in 2015; the design is inspired by Frank Lloyd Wright (found at http://www.tubemapcentral.com).

the *node–link metaphor* is commonly used: each individual datum is represented by a node and the connection between different data elements is modelled by connecting the nodes via arcs. In most of today's literature, nodes are called *vertices* and links are called *edges*, reusing the names established in discrete maths. Most commonly, heavy dots are used to represent vertices and edges are either straight-line segments or simple curves. Modifying and augmenting these objects already allows for a lot of different drawing styles: Is the vertex labeled? How is it shaped? Are different colors used to represent different data types? Are the edges curvy or straight? Do they have arrow-tips indicating directions? Is additional data – such as distance or connection strength – represented? An extensive overview on different drawing styles is given in *Semiology of graphics* by Jacques Bertin [Ber83]. Of course, graphic standards are not limited to changing the visual representations of vertices and edges.

This work focuses on graphic standards that restrict where vertices are allowed to be placed. In classic graph drawing literature, the goal is often to draw a given graph onto a two-dimensional plane – like a sheet of paper, blackboard, or computer monitor. The placement of each vertex is described by assigning two coordinates (usually using real numbers to represent x- and y-coordinate). We now list several important results on different graphic standards.

De Fraysseix, Pach, and Pollack [dFPP90], as well as Schnyder [Sch90], considered drawings of bounded size with straight-line edges and vertices placed at integer coordinates. They managed to show that every graph that can be drawn planar – without crossing edges – has a layout using the integer grid whose size is quadratic in the number of nodes.

Networks representing hierarchical structures, such as trees, can be drawn using the layered layout by Sugiyama, Tagawa, and Toda [STT81], where vertices of the same level obtain the same y-coordinate. A more general case of these layered layouts are radial layouts: Layers are represented by concentric circles and vertices on the same level have the same distance to the center. Outerplanar graphs can be imagined as a case: All vertices have to be placed on the same circle, all edges are routed inside the circle and no pair of edges is allowed to intersect.

Outerplanar graphs are also known as planar permutation graphs, introduced by Chartrand and Harary [CH67]. Fruchterman and Reingold [FR91] take a different approach, by aiming to restrict how closely together vertices are allowed to be drawn and aiming for an equal edge length. They model the edge lengths using repulsive forces of springs between vertices, trying to find an equilibrium by expanding dense vertex clusters at the expense of less dense parts.

Another layout restriction concerning polyline edges focuses on limiting the number of different edge slopes used in the drawing. Rectilinear and octilinear drawings – used in graph layouts such as Manhattan-geodesic drawings (for example, see Katz, Krug, Rutter, and Wolff [KKRW10]) and classic metro-map drawings (see Nöllenburg and Wolff [NW11]) – are well-studied graphic standards with a large set of applications. For a broad overview and diverse selection of results on bounded slope numbers, we refer to the work of Dujmović, Suderman, and Wood [DSW07].

Unfortunately, as usual in algorithmic fields, there are many graph drawing and layout problems that cannot be solved efficiently unless we have $\mathcal{P} = \mathcal{NP}$. Some interesting examples are listed below.

Eades and Wormald [EW90] showed that for a given graph G with prescribed edge lengths it is $\mathcal{NP}$-hard to determine whether there is a crossing-free drawing of G in which the edges are straight-line segments of the prescribed lengths.

As shown by Argyriou, Bekos, and Symvonis [ABS12], it is also $\mathcal{NP}$-hard to decide for a given non-planar graph, whether the vertices can be placed such that (a) all edges are straight-line segments and (b) if two edges cross, they do so at right angles. Minimizing the total area needed to draw a planar graph using straight-line edges at integer coordinates is $\mathcal{NP}$-hard, shown by Krug and Wagner [KW07]. Cabello [Cab06] showed that even deciding for a given set of points and a planar graph, if there is a mapping of the vertices to the coordinates such that the resulting drawing is planar is $\mathcal{NP}$-hard.

As discussed above, we consider a drawing to be good if it is visually pleasing and transports information without ambiguity. Some graphic standards are better suited for placing labels next to vertices, some allow for a better perception of graph distance and connectivity inside the network. While we have hinted at applications in transit map drawing and organizational charts, this work puts the focus on theoretical results, broadening the understanding of what makes a layout problem hard or easy.

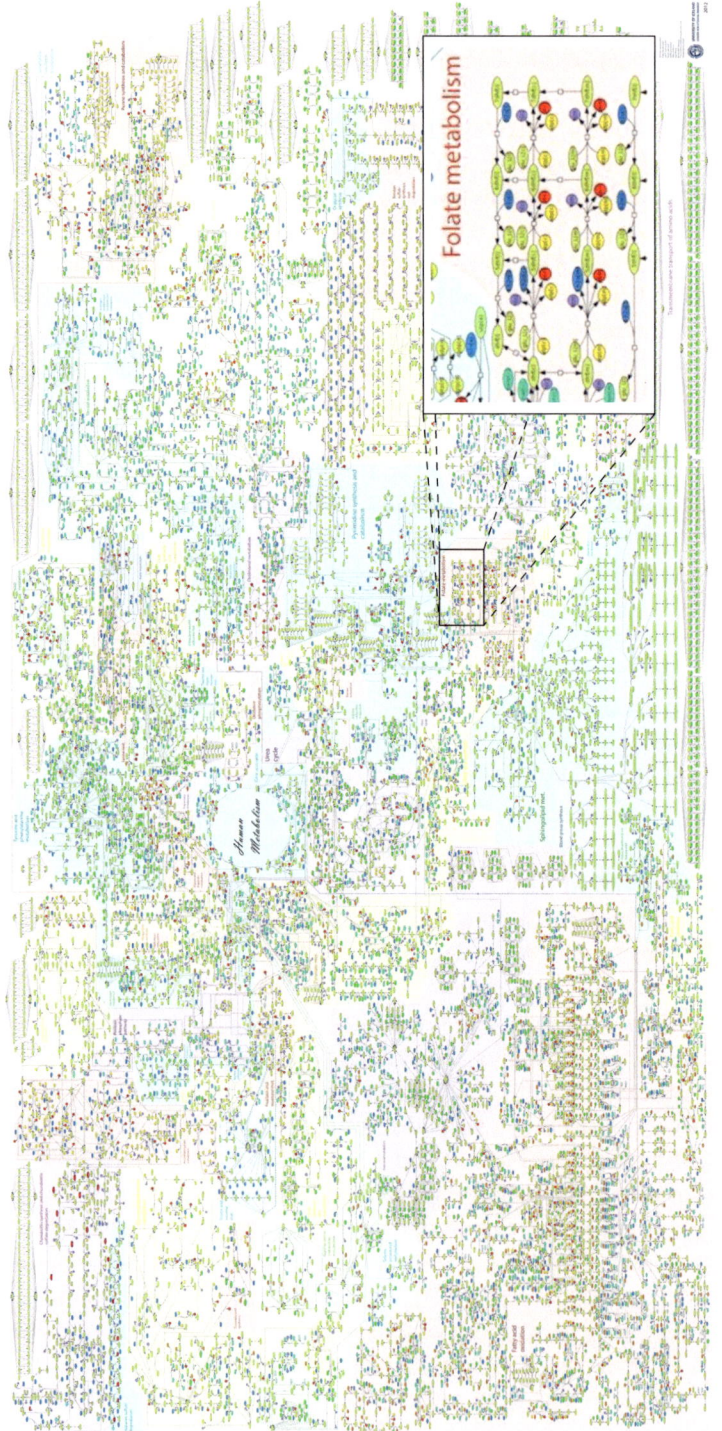

Figure 1.2: The Human Metabolism Map, published by Thiele et al. [TSF⁺13], is the largest hand-drawn network in biology. An interactive online explorer is available at https://www.vmh.life/#reconmap, see Noronha et al. [NDG⁺16].

Outline of this Book

In this work, we answer selected questions on two different types of graph layout strategies, each of which restrict vertex placement. This naturally partitions the chapters of this book into two parts as follows:

Part One (Chapters 3 and 4) examines graph drawings that have all vertices mapped to the boundary of a common (outer) face. An example for drawings in this layout style can be found in Figure 1.3 (b). Part Two (Chapters 5, 6, and 7) looks into graph drawings with vertices at integer coordinates. Refer to Figure 1.3 (c) for an example.

Additionally, Chapter 2 reviews the required basics in graph theory, algorithms, and computational complexity. And finally, Chapter 8 summarizes the contents of this book, giving an outlook by pointing to several open research questions.

Part One: Vertices on a Common Outer Face

Beyond Outerplanarity. In the first chapter of Part One, we look into the structural properties of graphs that admit non-planar *convex drawings*; that is, all vertices are placed in convex position – defining the boundary of the outer face – and all edges are straight lines going between the vertices.

In Chapter 3, we consider two families of graph classes with nice convex drawings. These families are defined by the crossing patterns that edges of members of these classes are allowed to make. These classes are *outer k-planar* graphs – where each edge is crossed by at most k other edges – and *outer k-quasi-planar* graphs – where no k edges can mutually cross.

For the family of outer k-planar graphs, we show $(\lfloor\sqrt{4k+1}\rfloor+1)$-degeneracy. As an immediate consequence we get that every outer k-planar graph can be $(\lfloor\sqrt{4k+1}\rfloor+2)$-colored, and this bound is tight. We further show that every outer k-planar graph has a balanced separator of size at most $2k+3$. This implies that the treewidth of such graphs is $O(k)$. For each fixed k, these small balanced separators allow us to test outer k-planarity in quasi-polynomial time, hence recognizing membership in any class of this family is not $\mathcal{NP}$-hard unless the Exponential Time Hypothesis fails.[1]

The other family of graph classes considered in this chapter is that of outer k-quasi-planar graphs. We discuss the edge-maximal graphs of this family, which have been considered previously under different names (such as chords of a convex polygon by Capoyeleas and Pach [CP92]). We also construct planar 3-trees that are not outer 3-quasi-planar, showing that planar graphs and outer 3-quasi-planar graphs are incomparable.

In the last section of this chapter, we further restrict outer k-planar and outer k-quasi-planar drawings to *closed* drawings and subsequently to *full* drawings. A drawing of a graph from either family is closed when the sequence in which the vertices appear on the outer face's boundary is a cycle in the graph; a drawing of a graph from either family is full, when no crossing occurs on the boundary of the outer face. Naturally, any closed

---

[1] Details on the Exponential Time Hypothesis can be found in the work of Impagliazzo and Paturi [IP01].

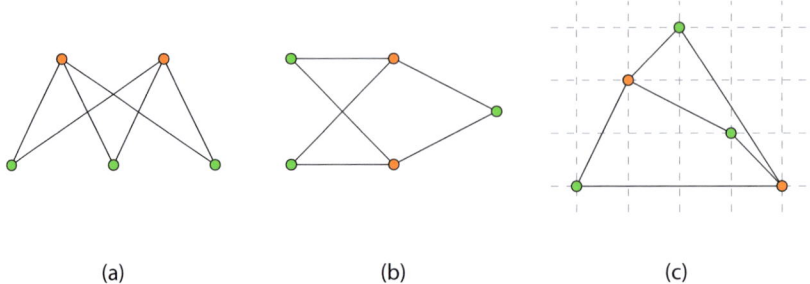

(a)   (b)   (c)

Figure 1.3: Illustration of considered layout styes. Three drawings of the complete bipartite graph $K_{2,3}$: (a) A drawing with opposing subsets, (b) an outer 1-planar drawing that is also outer 3-quasi-planar, (c) and a crossing-free triangular-shaped grid drawing.

drawing is also full. For each k, we express *closed outer k-planarity* and *closed outer k-quasi-planarity* in *extended Monadic Second-Order Logic*. Thus, by Courcelle's Theorem [Cou90], for each fixed k, closed outer k-planarity is linear-time testable since outer k-planar graphs have bounded treewidth. Using the test for closed outer k-planarity as a subroutine, we can also test full outer k-planarity in linear time.

    This chapter is based on joint work with Steven Chaplick, Myroslav Kryven, Giuseppe Liotta, and Alexander Wolff [CKL+17].

Polygonal Boundaries. Many drawing algorithms for planar graphs work recursively. Given a planar drawing of a subgraph, each step of the recursion extends it. The extensions work in a way that maintains the structural properties of the previous drawing that made the extension possible. This is oftentimes done by joining subgraphs at distinct vertices – e.g. combining two trees by merging the root vertex of one tree with a leaf vertex of the other – or by drawing subgraphs inside the faces of the previous drawing, connecting the new vertices to those defining the face. For the latter, the shape of the face to be drawn in can be prescribed, requiring the extended drawing to respect this shape without inducing edge crossings. This raises the question if a given planar graph can be drawn inside a prescribed outer face such that the resulting drawing is also crossing free.

    In Chapter 4, we consider this question in a special setting: All vertices of the subgraph are already placed on the prescribed face's boundary. Thus drawing the subgraph only requires adding the missing edges inside the face. If the shape of the face is convex, this problem is equivalent to testing whether the subgraph with given embedding is outerplanar. If it is not convex, insisting on drawing edges as line-segments can easily make them cross the face's boundary. Therefore we consider the following problem: Given a drawing of the outer face as a simple polygon with p corners, an outerplanar graph with n vertices and a mapping of the vertices to the face's boundary, can the edges be drawn inside the face with at most one bend per edge? We prove that this can be decided in $O(pn)$ time. We do so by giving an algorithm that, in the positive case, also outputs such a drawing.

This chapter is based on joint work with Patrizio Angelini, Philipp Kindermann, Lena Schlipf, and Antonios Symvonis [AKL+20].

Part Two: Vertices at Integer Coordinates

Moving to the Grid Optimally. Until this point we were concerned with where the vertices are placed, but not at what precision this placement is stored. Taken as a given that real computers work within hard limitations, and can only store numbers at finite precision, approximating and rounding numbers in the process is inevitable. Greene and Yao [GY86] noted that repeatedly performing geometric operations – such as checking point-in-area containment and line intersection – on coordinates of finite precision can result in rapidly growing rounding errors.[2]

Actually working with limited resources creates the need for an algorithm that transforms a given drawing into a drawing of lower coordinate precision without damaging the embedding or completely losing geometric similarity. Preserving topology and geometric similarity are most important when working with data like real-world road networks. The transformation also removes unnecessary detail and reduces space consumption as well as the computation time of algorithms working with the coordinates. In the first chapter of Part Two – Chapter 5 – we investigate the TOPOLOGICALLY-SAFE GRID REPRESENTATION problem for given straight-line drawings of planar graphs.[3] We show that the problem is $\mathcal{NP}$-hard for several different objective functions and provide an integer linear programming formulation to compute optimal solutions. We also provide an experimental evaluation on the performance and limitations of our program.

This chapter is based on a Master's Thesis as well as joint work with Thomas C. van Dijk and Alexander Wolff [Löf16, LvDW16].

Rounding to the Grid Heuristically. We originally started looking into the TOPOLOGICALLY-SAFE GRID REPRESENTATION problem with a geographic application in mind, engaging it from a graph-drawing perspective. Chapter 5 presents an $\mathcal{NP}$-hardness result and an exact ILP formulation that is infeasible for practical applications, so we ask for an efficient algorithm that transforms a given drawing into a topologically equivalent grid drawing.

In Chapter 6, we tackle this problem from a different angle by providing a randomized heuristic algorithm. Our algorithm consists of two stages of simulated annealing, each with a different objective function. Stage One focuses on finding a feasible solution

---

[2] Motivated by actually putting the drawings onto the *screen*, work on this topic usually considers points to be centered inside pixels. Transforming pixel-centers to (crossing) points on the integer grid only requires uniformly shifting all object by half of a unit in both dimensions.

[3] Most of the results in Chapter 5 have already been published as part of a Master's Thesis [Löf16]. We choose to include them here again for two reasons: To provide a more profound experimental evaluation; and as a foundation for the work presented in Chapter 6.

– a non-optimal drawing with all vertices placed on grid points – by reducing the overall "density" of the drawing, moving vertices away from each other. Stage Two takes the
feasible drawing, but switches the objective to reducing the total movement of vertices
induced by the first stage. We discuss various feasibility procedures and evaluate their
applicability on geographic networks. We demonstrate that a straightforward annealing
approach without the first stage has difficulty finding any feasible solution at all. We
also discuss parameter selection for the second simulated annealing step, which tries to
reduce rounding cost of the drawing.

This chapter is based on joint work with Thomas C. van Dijk [vDL19].

Recognizing Nets of Orthogonal Polyhedra. In computational geometry, the graph
defined by the corners and creases of a polyhedral surface is a commonly studied object
– often called the skeleton. The Rigidity Theorem given by Cauchy in 1813 can be restated
within today's notation as follows: When a graph is the skeleton of the surface of a convex
polyhedron and the angles within each face of that graph are given, the dihedral angles
of the faces on the surface are uniquely determined.

This naturally raises questions about other polyhedral surfaces to which graphs can
be embedded uniquely. Biedl and Genç [BG09] studied orthogonal polyhedral surfaces
of genus 0 with connected graphs – that is, the angles between edges as well as those
between faces are multiples of 90° that are topologically equivalent to the surface of the
sphere S^2 in $\mathbb{R}^3$. Restricting faces to be orthogonal polygons and requiring edge lengths
to be integer, we get that all corners and creases of the polyhedral surface end up being placed on the three-dimensional integer grid. Biedl and Genç give the linear-time
BUNDLEORIENTATION algorithm to translate Cauchy's theorem to orthogonal polyhedral
surfaces of genus 0: This algorithm can determine the unique set of dihedral angles for
a given connected graph with orthogonal faces or report that no such set of angles exists. Biedl and Genç also showed that testing realizability is $\mathcal{NP}$-hard when the graph is
disconnected.

In Chapter 7, we consider an essential question that was left open by Biedl and Genç –
whether or not a similar translation exists for orthogonal polyhedral surfaces of genus 1
or higher. They give an example instance of genus 1 on which their algorithm can fail. To
answer their open question in the affirmative, we introduce the ITERATEDBUNDLECOL-
ORING algorithm. It uses the original BUNDLEORIENTATION algorithm repeatedly and
exhaustively. We show that it is capable of finding a set of dihedral angles for orthogonal polyhedral surfaces of arbitrary genus. We do so by arguing how it re-discovers the
dihedral angles matching those of a polyhedron realizing the input graph.

This chapter is joint work with Steven Chaplick and Thomas C. van Dijk.

Chapter 2
Basic Definitions

This book considers questions about graph drawings under different layout constraints. To answer them, we provide algorithms as well as results from complexity theory. For a general overview on these topics, we refer to two standard textbooks – *Introduction to Algorithms* by Cormen, Leiserson, Rivest, and Stein [CLRS13] for algorithms and *Computers and Intractability: A Guide to the Theory of NP-Completeness* by Garey and Johnson [GJ79] for complexity theory.

In this chapter, we define the essential concepts used throughout this book. First, we discuss the basics of graphs, including notation and terminology commonly used in graph theory and graph drawing. Then, we give the basics of algorithms, covering asymptotic runtime notation and standard considerations from computational complexity.

2.1 Graphs

2.1.1 Combinatorics

Graph Terminology. A *graph* $G = (V, E)$ is a tuple of sets V and E; the elements of V are the *vertices* – or *nodes*[1] – of G, the elements of E are the *edges*. We use $n = |V|$ and $m = |E|$ to refer to the sizes of these subsets respectively. In an *undirected* graph, each edge $e \in E$ is an unordered pair $\{u, v\} \in V \times V$ of vertices, whereas *directed* graphs use ordered tuples $(u, v) \in V \times V$, indicating that the edge goes from start vertex u to end vertex v. We use $V(G)$ and $E(G)$ to address the sets of vertices and edges in G respectively whenever the graph in question is not immediately clear from context.

The vertices of an edge are also called its *endpoints*. In this thesis, we do not treat directed edges differently; hence, we use the tuple-notation for both variants, considering all edges to be undirected. Whenever the direction of an edge is relevant, we explicitly state the direction the edge is going. We say that a graph G is *connected* when for every non-empty proper subset of vertices $A \subset V(G)$, we find at least one edge $(u, v) \in E(G)$ with $u \in A$ and $v \in V \setminus A$.

For an edge $e = (u, v)$, we say that u and v are *incident* to e and vice versa. In addition, two vertices are *adjacent*, if they are connected by an edge. Two edges are *incident*, if they share a common endpoint. The *degree* $\deg v$ of a vertex v is the number of edges incident to v.

---

[1] We rarely use the term "node". We usually use "node" to refer to vertices in auxiliary graphs, making sure they are not mistaken for elements of the primary graph.

A *subgraph* $G' = (V', E')$ of a graph G has the following properties: The vertex set $V' \subseteq V(G)$ is a subset of the original vertex set and the edge set $E' \subseteq E(G)$ is a subset of all edges $(u, v) \in E(G)$ with $u, v \in V'$. If a graph is not connected, the maximal connected subgraphs of G are called its *connected components*. Naturally, each connected component itself is a graph – possibly containing only as little as a single isolated vertex.

The *subdivision* of an edge $e = (u, v)$ is created by deleting e from G, adding another vertex v' and adding the edges (u, v') and (v', v). A graph G' is a *subdivision* of graph G if G' can be created subdividing some of the edges of G, or if G' is a subdivision of a subdivision of G.

Special Graphs and Graph Properties. Next, we look into graph classes defined by structural properties of abstract graphs. In the following, let $G = (V, E)$ be a graph.

Let $u, w \in V$ be vertices of G. We say that G contains a *path* P of length k connecting u and w if there is a k-element subset of edges $\{e_1, e_2, \ldots, e_k\} \subset E(G)$ creating the sequence $e_1 = (u, v_1), e_2 = (v_1, v_2), e_3 = (v_2, v_3), \ldots e_k = (v_{k-1}, w)$. We call u and w the end vertices of the path. Each end vertex appears only once in the edge sequence, all other vertices appear exactly twice. Naturally, we have $|V(P)| = |E(P)| + 1$. A *cycle* of length $k \geq 3$ is a closed path[2] – a sequence of edges starting and ending on the same vertex, that is, $e_k = (v_{k-1}, u)$ and $e_1 = (u, v_1)$. A *chord* is an edge e connecting two vertices on the same cycle (of length at least 4) that e itself is not part of[3]. A cycle is an *induced cycle* if no pair of vertices on that cycle is connected by a chord.

A *tree* is a connected graph, that does not have a subgraph that is a cycle. The degree-1 vertices of a tree are called its *leaves*. A tree containing all vertices of a given graph G is a *spanning tree* of G. A *Hamiltonian path* is a path of G that visits every vertex of G exactly once. A *Hamiltonian cycle* is a *Hamiltonian path* that is also closed. When G has a Hamiltonian cycle, we call G a *Hamiltonian* graph. The edges highlighted in red in Figure 2.1 (a) are a Hamiltonian cycle of that graph.

The edge set E of the *complete graph* $K_n = (V, E)$ on n vertices contains all possible two-element subsets of V, that is, $E = \binom{V}{2}$ (and $m = n \cdot (n-1)$). A graph contains a *clique* of size k when it has K_k as a subgraph. The complete graph K_5 is shown in Figure 2.1 (a).

A *(vertex) coloring* of G is a function $c: V \to C$ mapping the vertices of G to a fixed set of colors[4] C such that we have $c(u) \neq c(v)$ for every edge $(u, v) \in E$. The *chromatic number* $\chi(G)$ is the size of the smallest set C_{min} for which a valid coloring of G exists; for simplicity we say that a graph G with $\chi(G) = x$ is *x-colorable*. Naturally, the chromatic number of the complete graph is $\chi(K_n) = n$, and a famous result by Appel and Haken [AH76] states that all planar graphs are 4-colorable; trees, paths, cycles of even length are 2-colorable, non-trivial cycles of odd length require a third color. A valid 5-coloring of K_5 is shown in Figure 2.1 (a).

---

[2] Cycles of length 1 are edges with both end points being the same vertex, usually called (self-) *loops*. Cycles of length 2 can only appear in directed graphs or multigraphs; they are usually called *lenses*. Both special cases will not be considered in this thesis.

[3] Think of a chord as a shortcut through the cycle. Naturally, triangles cannot have shortcuts.

[4] Despite the name, we generally use natural numbers or letters as colors, not "red, blue, green, …".

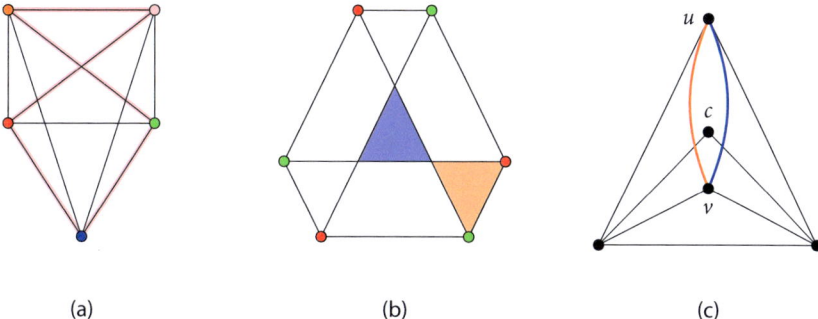

(a)   (b)   (c)

Figure 2.1: Illustrations of basic definitions: (a) The complete graph K_5: One of many Hamiltonian cycles in red; the vertex-colors show a valid 5-coloring. (b) The complete bipartite graph $K_{3,3}$: The central (blue) face is bounded only by partial edges and thus has no vertices incident to it; the bottom-right (orange) is not adjacent to the central face. (c) The edge (u, v) can be drawn in two different ways – going around vertex c to the left (orange drawing) or to the right (blue drawing). Each drawing crosses a different edge incident to c, creating two different subdivisions that both follow the same rotation system.

The 2-colorable graphs are also called the *bipartite* graphs, since any valid 2-coloring partitions their vertex set into two groups V_ℓ and V_r. The special family of *complete bipartite* graphs contains the graphs $K_{a,b}$ with $a = |V_\ell|$ and $b = |V_r|$. They contain every possible edge between these groups – we have $E = V_\ell \times V_r$.

Recognition.   When considering a given structural property, one would generally want to test whether a given graph admits it. The *recognition problem* can generally be stated as follows: "Does graph G belong to the class of graphs with property X?" Some recognition problems are easy – such as testing planarity –, while testing other properties can be very challenging – like searching for a Hamiltonian path. Oftentimes, in graph drawing – as well as in this book – the property to test for is whether or not a given graph can be drawn according to a specific graphic standard.

2.1.2   Drawings

In the previous section, we have defined the elements of abstract graphs as well as some of their structural properties that are of general relevance to this book. To visually represent – or *draw* – an abstract graph, we use the *node-link metaphor*. We have already worked with the intuitive nature of this metaphor in the introduction when giving an overview on different graph layout approaches. We now formally define the basic concepts and notation used for drawing graphs using this metaphor.

Vertices and Edges. A drawing Γ of graph G is a visual representation placed in some k-dimensional Euclidean space. The drawing maps each vertex v of G to a k-dimensional *coordinate* vector $\Gamma(v)$ with entries from $\mathbb{R}$. Since we want to visualize drawings – to actually be able to look at them –, we restrict ourselves to either $k = 2$ or $k = 3$. For $k = 2$, we get the well-known *Cartesian plane* $\mathbb{R}^2$, most commonly associated with a blackboard or sheet of paper to draw on; for $k = 3$, we get the 3-dimensional *space* $\mathbb{R}^3$. Producing a 3-dimensional drawing oftentimes involves drawing onto the surface of some 3-dimensional object, physically building objects representing the vertices and edges of G or using $3D$ computer graphic tools.[5] We identify the vector $\Gamma(v)$ with its associated vertex and refer to the entries of $\Gamma(v)$ as the x-, y- (and z-) coordinates. Vertices are commonly represented by "heavy" dots, as in Figure 2.1.

Edges are drawn using *Jordan curves*. A Jordan curve is a continuous injective map d of the interval $[0, 1]$ to $\mathbb{R}^2$ (or $\mathbb{R}^3$ respectively). That is, for $i, j \in [0, 1]$ we have the following two conditions: When j converges towards i, we get $\lim_{j \to i} d(j) = d(i)$ – there are no abrupt changes in value – and for $i \neq j$ we have for $d(i) \neq d(j)$ – a Jordan curve never occupies the same point in space twice. A drawing $\Gamma(e)$ of an edge $e = (u, v)$ is a Jordan curve where one endpoint is $\Gamma(u)$ and the other endpoint is $\Gamma(v)$. If the map is also linear, we call the resulting curve a *straight-line segment*. For simplicity, we identify an edge e and its drawing $\Gamma(e)$.

Planarity and Beyond. In a drawing, the curves of two edges e and f can cross at some interior point– that is, we get $\Gamma(e) \cap \Gamma(f) \neq \emptyset$ [6]; if they share interior points, we call each such point a *crossing*. If two edges cross, we insist that they do so at interior points, that is, we do not allow for an endpoint of one edge to b an interior point of another edge. A drawing in the plane without crossing edges is a *planar* drawing and a graph that has some planar drawing is called a *planar* graph; if a graph has no planar drawing, it is nonplanar. A famous result by Fáry [Fár48] states that every planar graph also has a planar straight-line drawing. Planarity can also be characterized combinatorially. Tutte [Tut63] states *Kuratowski's* Theorem as follows: A graph is planar if and only if none of its subgraphs is a subdivision of the complete graph K_5 or the complete bipartite graph $K_{3,3}$. These graphs are also known as the Kuratowski graphs; they are shown in Figure 2.1 (a) and (b) respectively.

The notion of planarity can be relaxed as follows: For any constant k, a graph is k-*planar*, if it has a drawing in the plane in which no edge participates in more than k crossings; if a graph is k-planar, it is also $(k + 1)$-planar. Notice that Fáry's theorem does not generalize to 1-planarity. A more detailed overview on beyond-planar graph classes can be found in Chapter 3.

---

[5] During our research for Chapter 7, we used all of these techniques. We built polyhedral surfaces from paper cut-outs and plastic tiles, then drawing vertices and edges onto them. Several of the figures found in Chapter 7 are screenshots of digital $3D$ models created using our own implementation in JavaScript.

[6] This does not include incident edges "crossing" at their shared endpoint.

Rotation Systems and Embeddings. The *rotation system* of a drawing of a graph is defined as the set of cyclic orders of incident edges around each vertex, usually considered consistently counterclockwise. The edges of a drawing in the plane[7] subdivide that plane into disjoint regions called *faces*. In the plane, there is exactly one unbounded *outer* face and (possibly) some *inner* faces. The boundary of each inner face is a closed curve composed of a cyclic sequence of edges connecting incident vertices and *partial* edges connecting vertices and/or crossings. Notice that an edge can appear twice in this sequence. We say that an edge or vertex is *incident* to a face if it belongs to the closed curve bounding it. Similar to above, two faces are adjacent if their boundaries share an edge. An *embedding* of a graph is a rotation system together with a prescribed outer face. The subdivision created by a drawing of graph G also induces one embedding of G but for a given embedding, there is an infinite number of drawings realizing it.

We can also use a rotation system to describe an embedding. For planar graphs, these definitions are equivalent; for nonplanar graphs, there can be multiple different subdivisions – creating faces bounded by combinatorially different curves – following the same rotation system (see Figure 2.1 (c)). We will sometimes prescribe the combinatorial embedding of a graph to then try finding a drawing matching the given embedding.

2.2 Algorithms

In this section, we define the algorithmic concepts relevant to this book. An *algorithm* is a finite sequence of well-defined instructions, designed to be executed by some machine (such as Turing machines, computers, or humans). An algorithm is *deterministic*, if when given an input, it will always perform the same sequence of steps to arrive at the same output. A *non-deterministic* algorithm has some "freedom of choice" when performing the next instruction – some choices might lead to different outputs than others. A *randomized* algorithm employs some random number generator – such as flipping coins or creating a random order of the elements of a set. The computational model used throughout this book is the *real RAM model* – a hypothetical machine capable of infinite-precision arithmetic operations on real numbers in constant time. Details on this model can be found in the book *Computational Geometry* by Preparata and Shamos [PS85].

2.2.1 Asymptotic Runtime

When looking at families of graphs or the running time of algorithms, one can often recognize some *asymptotic* behaviour. We describe such behaviours using *Landau symbols* – also known as *Big Oh Notation*. Let $f : \mathbb{N} \to \mathbb{N}$ be a function that maps a natural number to a natural number – e.g. input size and running time for an algorithm or number of vertices (usually called n) and edges (m) for some graph. We define the following classes of functions with respect to function f:

---

[7] Or on some other two-dimensional surface.

$$O(f) = \{g\colon \mathbb{N} \to \mathbb{N} \mid \exists c > 0, \quad \exists n_0 > 0, \forall n > n_0\colon \quad 0 < f(n) \le c \cdot g(n) \}$$
$$\Omega(f) = \{g\colon \mathbb{N} \to \mathbb{N} \mid \exists c > 0, \quad \exists n_0 > 0, \forall n > n_0\colon c \cdot g(n) \le f(n) \}$$
$$\Theta(f) = \{g\colon \mathbb{N} \to \mathbb{N} \mid \exists c_1, c_2 > 0, \exists n_0 > 0, \forall n > n_0\colon c_1 \cdot g(n) \le f(n) \le c_2 \cdot g(n) \}$$

The growth of a function g that *lies in the order of* f – or for short $g \in O(f)$ – is upper-bounded by that of f. Symmetrically, we have that for $g \in \Omega(f)$, the growth of g is lower-bounded by that of f. Intuitively, we have that $g \in \Theta(f)$ implies that f and g grow alike – that is, $g \in O(f)$ and $g \in \Omega(f)$, thus f upper- and lower-bounds g for different constants.

If we have $f \in O(n^k)$ for some constant k, we say that f is a *polynomial* function; an algorithm with runtime $O(n^k)$ is an *efficient* or *polynomial-time* algorithm. Opposed to efficient algorithms are the *exponential-time* algorithms. An exponential algorithm has a runtime that is lower-bounded by some exponential function (a function $f(n) = c^n$ in which the parameter appears in the exponent at least once and with some base $c > 1$). Between polynomial and exponential functions are the *quasi-polynomial* functions. A function f is quasi-polynomial, if we have $f \in O(2^{(\log n)^c})$ for some constant $c > 0$.[8] A *quasi-polynomial-time* algorithm has a runtime that can be bounded using a quasi-polynomial function.

2.2.2 Complexity and Hardness

Complexity Classes. There are two different kinds of problems that we discuss in this book. For *decision* problems, we try to find a yes/no-answer; whereas for *optimization* problems, we look for the best answer – optimizing some *objective* function.

The complexity class $\mathcal{P}$ is the class of all decision problems that are solvable by some efficient deterministic algorithm. Similarly, the class $\mathcal{QP}$ is the class of all decision problems solvable by quasi-polynomial-time algorithms. On the other hand, the class $\mathcal{NP}$ is the class of problems solvable by non-deterministic polynomial-time algorithms. We clearly have that $\mathcal{P} \subset \mathcal{NP}$, but the question whether the two classes are the same or not – $\mathcal{P} = \mathcal{NP}$ or $\mathcal{P} \subsetneq \mathcal{NP}$ – remains open. From an algorithm standpoint, most results are stated under the standard assumption that $\mathcal{P} \ne \mathcal{NP}$.[9]

A problem T is $\mathcal{NP}$-*hard*, when there is a polynomial-time reduction from every problem $R \in \mathcal{NP}$ to T. A *reduction* from R to T is a transformation that translates an instance I_R for problem R into an instance I_T for problem T such that there is a valid solution to I_R if and only if there is one to I_T. A problem is $\mathcal{NP}$-*complete* if it is $\mathcal{NP}$-hard and in $\mathcal{NP}$.

---

[8] For $c = 1$ we get the linear functions, whereas for $c < 0$ we get *sub-polynomial* functions.
[9] This question has been open for about 60 years. In 2012, William Gasarch [Gas12] conducted a survey among computer scientists. Out of 152 participants, 83% expected the two classes to not be equal.

Boolean Satisfiability.   Finally, we define one of the classic problems shown to be $\mathcal{NP}$-complete by Richard Karp [Kar72] in 1972 – namely *Satisfiability with at most 3 literals per clause*, also known as 3SAT. Cook [Coo71] and Levin [Lev73] independently considered the *Boolean Satisfiability Problem* (or SAT), showing that it is the first $\mathcal{NP}$-complete problem; this result is therefore known as the *Cook-Levin theorem*. An instance of SAT – a Boolean *formula* – consists of Boolean *variables* and the three operators AND, OR, and NOT (denoted by the symbols $\wedge$, $\vee$, and $\neg$ respectively). A formula is called *satisfiable* if it is true for some assignment of Boolean values to the variables. SAT is the problem of deciding whether a given formula is satisfiable.

The variables appear as negated or unnegated *literals* – such as $\neg x$ or x. Each clause is a disjunction of literals – for instance $(x \vee \neg y \vee z)$ – and a formula is in *conjunctive normal form*, if it is a conjunction of clauses. Any SAT formula can be translated to an equivalent formula in conjunctive normal form. The issue with this is that the translated formula's size can be exponentially larger than that of the original one.

The 3SAT problem is a special version of SAT; a 3SAT formula is in conjunctive normal form and in addition, every clause contains no more than three literals. While 3SAT is $\mathcal{NP}$-complete, the variant 2SAT – restricting the clauses to contain up to two literals each – can be solved efficiently.

Exponential Time Hypothesis.   The *Exponential Time Hypothesis (or ETH for short)* (as stated by Impagliazzo and Paturi [IP01]) is a complexity theoretic assumption defined as follows: Let $s_k = \inf\{\delta:\text{there is an } O(2^{\delta n})\text{-time algorithm to solve } k\text{SAT}\}$. The Exponential Time Hypothesis states that for $k \geq 3$ we have that $s_k > 0$; in other words, there is no sub-exponential-time algorithm – and subsequently also no quasi-polynomial-time algorithm – that solves 3SAT. Finding a problem that can be solved in quasi-polynomial time and that is also $\mathcal{NP}$-hard would contradict the ETH. In recent years, the Exponential Time Hypothesis has become a standard assumption from which many conditional lower bounds have been proven; Cygan et al. [CFK+15] provide a good summary on this topic.

Note that, in addition to violating the ETH, the existence of an $\mathcal{NP}$-hard problem which can be solved in quasi-polynomial time would also directly imply that *deterministic* and *nondeterministic exponential time* – $\mathcal{EXP}$ and $\mathcal{NEXP}$ – coincide. This can be proven by a padding argument similar to Proposition 2 by Buhrman and Homer [BH92]. Thus, having a quasi-polynomial algorithm for a problem implies that it is extremely unlikely for that problem to be $\mathcal{NP}$-hard. Thus, it is widely believed that $\mathcal{NP} \not\subseteq \mathcal{QP}$.

Approximability.   Similar to $\mathcal{NP}$-hard decision problems, there are also $\mathcal{NP}$-hard optimization problems. To any optimization problem, there is also a division variant – instead of asking for the best solution, one can also ask if there exists a solution above or below a certain threshold. The complexity class $\mathcal{NPO}$ contains all $\mathcal{NP}$-hard optimization problems.

One way to tackle a problem in $\mathcal{NPO}$ is to try finding an *approximation* algorithm – an algorithm with a provable guarantee on the solution quality. Let $OPT(I)$ be the optimal solution for an instance I of some optimization problem P and let $ALG(I)$ be the output of some approximation algorithm solving it. That algorithm is a *constant-factor* approximation, if there exists some constant c such that $c \cdot ALG(I) = OPT(I)$ holds for all instances I of problem P.

The class $\mathcal{APX}$ is the class of problems in $\mathcal{NPO}$ that have a polynomial-time constant-factor approximation algorithm.

Part I

Vertices on a Common Outer Face

Chapter 3

Outer *k*-Planar and
Outer *k*-Quasi-Planar Graphs

In the last decade, the focus in graph drawing has shifted from exploiting structural properties of planar graphs to addressing the question of how to produce well-structured and understandable drawings of general graphs.

This becomes even more important in the presence of edge crossings, giving rise to the topic of *beyond-planar* graph classes. The primary approach here has been to define and study graph classes which allow some edge crossings, but restrict these crossings in various ways. Two commonly studied such graph classes are:

1. *k-planar graphs*, the graphs which can be drawn so that each edge (Jordan curve) is crossed by at most *k* other edges.

2. *k-quasi-planar graphs*, the graphs which can be drawn so that no *k* pairwise non-incident edges mutually cross.

Following these definitions, the 0-planar graphs and 2-quasi-planar graphs are precisely the planar graphs. Additionally, the 3-quasi-planar graphs are simply called *quasi-planar*. Two highly relevant recent surveys on these classes are by Kobourov, Liotta, and Montecchiani [KLM17] and Didimo, Liotta, and Montecchiani [DLM19].

In this chapter we study these two families of classes of graphs under the restriction that the vertices are placed in convex position and edges mapped to line segments; i.e., we apply the above two generalizations of planar graphs to outerplanar graphs and study *outer k-planarity* and *outer k-quasi-planarity*.

Concepts.   In the following, we consider balanced separators, treewidth, degeneracy, coloring, edge density, and recognition to study these two classes. We start by defining the key concepts used in this chapter, then state our contribution to these beyond-outerplanar graph classes, to finally give an overview on the related work. We briefly define the most important graph theoretic concepts that we study in this chapter.

A graph is *d-degenerate* when every subgraph of it has a vertex of degree at most *d*. This concept was introduced by Lick and White [LW70] as a way to provide easy

---

A preliminary version of the contents of this chapter has appeared in the proceedings of Graph Drawing 2017 [CKL+17]. This is joint work with Steven Chaplick, Myroslav Kryven, Giuseppe Liotta, and Alexander Wolff.

bounds on the chromatic number. Namely, a d-degenerate graph can be $(d+1)$-colored by repeatedly removing a vertex of degree at most d. A graph class is d-degenerate if every graph in the class is d-degenerate. Note that the class of d-degenerate graphs is *hereditary*, that is, it is closed under taking subgraphs. Also note that outerplanar graphs are 2-degenerate, and planar graphs are 5-degenerate.

Given a graph $G = (V, E)$ with n vertices, a pair A, B of subsets of V is a *separation* of G if $A \cup B = V$, and no edge of G has one endpoint in $A \setminus B$ and the other in $B \setminus A$. The intersection $A \cap B$ is called a *separator* and the *size* of the separation (A, B) is $|A \cap B|$. A separation (A, B) of G is *balanced* if $|A \setminus B| \leq \frac{2n}{3}$ and $|B \setminus A| \leq \frac{2n}{3}$. The *separation number* of G is the smallest number s such that every subgraph of G has a balanced separation of size at most s. There is a polynomial relation between separation number and *treewidth* found by Robertson and Seymour [RS84]; namely, any graph with treewidth t has separation number at most $t + 1$ and, as Dvořák and Norin [DN19] recently showed, any graph with separation number s has treewidth at most $15s$. Graphs with bounded treewidth are of interest, among others, due to Courcelle's Theorem (see Theorem 3.12 [Cou90]), which implies that for graphs with bounded treewidth many problems can be solved efficiently.

3.1 Related Work and Contribution

Ringel [Rin65] was the first to consider k-planar graphs by showing that 1-planar graphs are 7-colorable. This was later improved to 6-colorable by Borodin [Bor84]. This is tight since K_6 is 1-planar. Many additional results on 1-planarity can be found in a recent survey paper by Kobourov, Liotta, and Montecchiani [KLM17]. Generally, each n-vertex k-planar graph has at most $O(n\sqrt{k})$ edges (shown by Pach [PT97]) and treewidth $O(\sqrt{kn})$ (shown by Dujmović, Eppstein, and Wood [DEW17]).

Outer k-planar graphs have been considered mostly for $k \in \{0, 1, 2\}$. Of course, the outer 0-planar graphs are the classic outerplanar graphs which are well-known to be 2-degenerate and have treewidth at most 2. It was shown by Babu, Khoury, and Newman [BKN16] that essentially every graph property is testable on outerplanar graphs. Outer 1-planar graphs are a simple subclass of planar graphs and can be recognized in linear time [ABB$^+$16, HEK$^+$14]. *Full outer 2-planar graphs* – a subclass of outer 2-planar graphs – can been recognized in linear time [HN16]. General outer k-planar graphs were considered by Binucci et al. [BGHL18]. Among other results, they showed that, for every k, there is a 2-tree which is not outer k-planar. Wood and Telle [WT07] considered a slight generalization of outer k-planar graphs in their work and showed that these graphs have treewidth $O(k)$.

The k-quasi-planar graphs have been studied extensively from the perspective of edge density. The goal here is to settle a conjecture of Pach, Shahrokhi, and Szegedy [PSS96] stating that every n-vertex k-quasi-planar graph has at most $c_k n$ edges, where c_k is a constant depending only on k. This conjecture has been settled in the affirmative by Ackerman and Tardos for $k = 3$ [AT07] and by Ackerman for $k = 4$ [Ack09]. The best known general upper bound is $(n \log n)2^{\alpha(n)^{c_k}}$, discovered by Fox, Pach, and Suk [FPS13], where

α is the inverse of the Ackermann function. Capoyleas and Pach [CP92] showed that any k-quasi-planar graph has at most $2(k-1)n - \binom{2k-1}{2}$ edges, and that there are k-quasi-planar graphs meeting this bound. More recently, it was shown by Geneson, Khovanova, and Tidor [GKT14] that the *semi-bar k-visibility graphs* are outer $(k+2)$-quasi-planar. However, the outer k-quasi-planar graph classes do not seem to have received much further attention.

The relationship between k-planar graphs and k-quasi-planar graphs was considered recently. While any k-planar graph is clearly $(k+2)$-quasi-planar, Angelini et al. [ABB$^+$20] showed that any k-planar graph is $(k+1)$-quasi-planar. More specially, it has also been shown that 2-planar graphs are also quasiplanar.

The *convex* (or *1-page book*) *crossing number* of a graph is the minimum number of crossings which occur in any convex drawing. This concept has been introduced several times (see Schaefer [Sch13] for more details). Masuda et al. [MKNF87] showed that determining the convex crossing number is $\mathcal{NP}$-complete. However, recently Bannister and Eppstein [BE18] used treewidth-based techniques (via extended Monadic Second-Order Logic – MSO$_2$) to show that the convex crossing number can be computed in linear FPT time, i.e., in $O(f(c) \cdot |I|)$ time where c is the convex crossing number and f is a computable function. Thus, for any k, the *outer k-crossing graphs* with n vertices and m edges can be recognized in time linear in $n + m$. Chaplick et al. [CvDK$^+$20] used MSO$_2$ to recognize graphs that have a circular layout with k bundled crossings[1].

Contribution. The rest of this chapter is structured as follows.

In Section 3.2, we consider outer k-planar graphs. We show that this graph class is $(\lfloor\sqrt{4k+1}\rfloor + 1)$-degenerate, and observe that the largest outer k-planar clique has size $(\lfloor\sqrt{4k+1}\rfloor + 2)$. This implies each outer k-planar graph can be $(\lfloor\sqrt{4k+1}\rfloor + 2)$-colored, and this is tight. We further show that every outer k-planar graph has separation number at most $2k+3$. For each fixed k, we use these balanced separators to obtain a quasi-polynomial time algorithm to test outer k-planarity, i.e., these recognition problems are not $\mathcal{NP}$-hard unless ETH fails.

In Section 3.3, we consider outer k-quasi-planar graphs. Specifically, we discuss the edge-maximal graphs which have been considered previously under different names, for example by Capoyleas and Pach [CP92], Dress, Koolen, and Moulton [DKM02], and Nakamigawa [Nak00]. We provide a novel approach to show that all edge-maximal outer k-quasi-planar graphs have the maximum number of edges, namely $2(k-1)n - \binom{2k-1}{2}$. We also relate outer k-quasi-planar graphs to planar graphs.

In Section 3.4, we restrict outer k-planar and outer k-quasi-planar drawings to *full* drawings (where no crossing appears on the boundary – see Figure 3.1 (a)), and to *closed* drawings (where the vertex sequence on the boundary is a cycle in the graph – see Figure 3.1 (c)). The case of full outer 2-planar graphs has been considered by Hong and Nagamochi [HN16]. They showed that full outer 2-planarity testing can be performed

---

[1] A *bundle* is a set of edges that travel in parallel. A *bundled crossing* occurs when two bundles cross – that is, every edge of one bundle crosses every edge of the other.

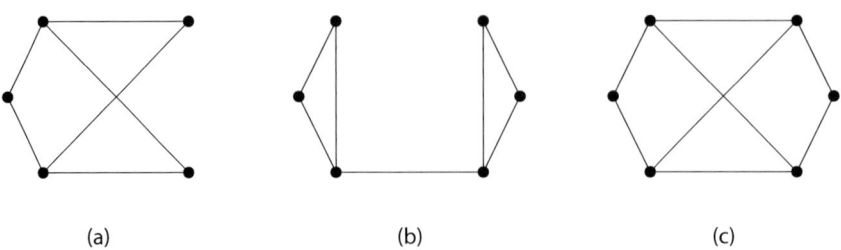

(a)   (b)   (c)

Figure 3.1: (a) An outer 1-planar embedding that is not full; (b) a full embedding that is not closed; (c) a closed outer 1-planar embedding.

in linear time. We first observe that a graph is full outer k-planar if and only if its maximal biconnected components are closed outer k-planar[2], and that this equivalence also holds for full outer k-quasi-planar graphs. Then, for each k, we express both *closed outer k-planarity* and *closed outer k-quasi-planarity* in *extended Monadic Second-Order Logic*. Thus since outer k-planar graphs have bounded treewidth, full outer k-planarity is testable in $O(f(k) \cdot n)$ time, for a computable function f. This result greatly generalizes the work of Hong and Nagamochi.

3.2   Outer k-Planar Graphs

In this section we show that every outer k-planar graph is $O(\sqrt{k})$-degenerate and has separation number $O(k)$. This provides tight bounds on the chromatic number, and allows for testing outer k-planarity in quasi-polynomial time.

3.2.1   Degeneracy and Coloring

We show that every outer k-planar graph has a vertex of degree at most $\sqrt{4k+1}+1$. First we note the size of the largest outer k-planar clique and then we prove that each outer k-planar graph has a vertex matching the clique's degree. This also tightly bounds the chromatic number in terms of k, i.e., Theorem 3.3 follows from Lemma 3.1 and Lemma 3.2.

Lemma 3.1. *Every outer k-planar clique has at most* $\lfloor\sqrt{4k+1}\rfloor + 2$ *vertices.*

Proof. Let a and b be two vertices of the largest clique. In any outer embedding the edge (a, b) partitions the other vertices into two sets S_ℓ and S_r – the vertices on the left and the right side of (a, b). Because all vertices are a clique, edge (a, b) is crossed $|S_\ell| \cdot |S_r|$ times. Therefore, the edge that is crossed the most has an almost equal number

---

[2] This was observed for full outer 2-planar graphs by Hong and Nagamochi [HN16].

of vertices on both sides; the total number of crossings then is $\left(\frac{n-2}{2}\right)^2 \leq k$ if n is even and $\frac{(n-3)(n-1)}{4} \leq k$ if n is odd. Therefore, for fixed k, the size of the clique is at most $\lfloor\sqrt{4k+1}\rfloor + 2$. $\qquad\square$

Lemma 3.2. *The largest possible minimum degree of a vertex in an outer k-planar graph G is $\sqrt{4k+1}+1$ and this bound is tight.*

Proof. In the following let δ be the largest minimum degree over all outer k-planar graphs and let G be some outer k-planar graph realizing δ. By Lemma 3.1 we know that G can contain a clique of size $\lfloor\sqrt{4k+1}\rfloor + 2$, so we have $\delta \geq \lfloor\sqrt{4k+1}\rfloor + 1$.

Hence, it remains to show that $\delta \leq \sqrt{4k+1}+1$. We use a contradiction argument to show that G cannot have minimum degree $\delta \geq \sqrt{4k+1}+2$ – it would then be too small to accommodate such a minimum degree vertex. We say an edge (a, b) *cuts* $h \in \mathbb{N}$ vertices, if there are h vertices to either the left or the right side of (a, b) with respect to the embedding of G.

Assume that edge (a, b) cuts h vertices of graph G and let $\text{cr}_{(a,b)}$ denote the number of edges crossing (a, b). Since we have minimum degree delta, we get the following:

$$\text{cr}_{(a,b)} \leq \delta h - h(h+1) \leq k. \tag{3.1}$$

Inequality (3.1) accounts for all edges incident to vertices cut by (a, b), subtracting all edges that can possibly go between those vertices or from any such vertex to a or b, as those edges cannot cross (a, b). Assume for contradiction that δ is larger, namely $\delta \geq \sqrt{4k+1}+2$; There is some number h^* of vertices that edge (a, b) would have to cut to violate k-planarity.

$$\text{cr}_{(a,b)} \geq \delta h^* - h^*(h^* + 1) \geq k + 1. \tag{3.2}$$

From Inequality (3.2) we get $h^* \geq \frac{1}{2}(\delta - 1 - \sqrt{(\delta-1)^2 - 4(k+1)})$. Consider the smallest such h^*. Further assume that no edge (a, b) cuts between h^* and h vertices inclusive. Thus, for edge (a, b) we get the following relationship:

$$\text{cr}_{(a,b)} \leq \delta h - h(h+1) + 2\left(\sum_{j=1}^{h-h^*} j\right). \tag{3.3}$$

The last term on the right side of Inequality (3.3) accounts for the absent edges that cut more than $h - h^*$ vertices. Now, assuming that (a, b) cuts $h+1$ vertices and we have $\delta > 2h^*$, we get the following from Inequality (3.2):

$$\begin{aligned}\text{cr}_{(a,b)} &\geq \delta h - h(h+1) + 2(\textstyle\sum_{j=1}^{h-h^*} j) + \delta - 2(h+1) + 2(h - h^* + 1) \\ &> k + \delta - 2(h+1) + 2(h - h^* + 1) > k.\end{aligned} \tag{3.4}$$

Inequality (3.4) is trivially satisfied for $\delta > \sqrt{4k+1}+2$. Therefore, with minimum degree $\delta \geq \sqrt{4k+1}+2$, there cannot be an edge that cuts more then $h^* < \frac{1}{2}\sqrt{4k+1}$

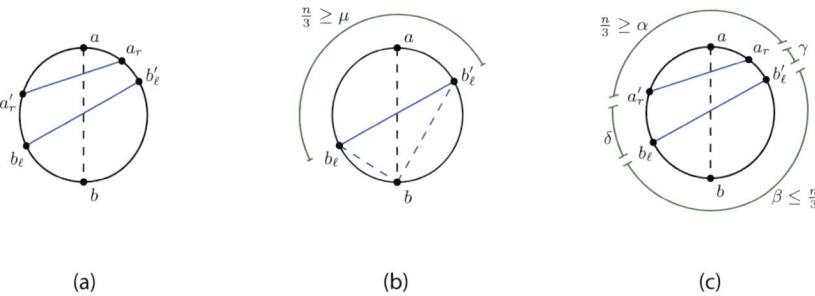

Figure 3.2: Illustrations for finding a balanced separator as in the proof of Theorem 3.4: (a) the pair of parallel edges (b_ℓ, b'_ℓ) and (a_r, a'_r); (b) case 1 in the proof; (c) case 2.

vertices in any outer k-planar graph. On the other hand, such a graph can have at most $2h^* < \sqrt{4k+1}$ vertices, which is not enough "other endpoints" to accommodate the minimum degree vertex required; a contradiction. $\qquad\square$

Having established that outer k-planar graphs are $(\sqrt{4k+1}+1)$-degenerate, we easily obtain the following result.

Theorem 3.3. *Each outer k-planar graph is $\sqrt{4k+1}+2$ colorable and this is tight.*

3.2.2 Quasi-polynomial time recognition via Balanced Separators

We show that outer k-planar graphs have separation number at most $2k+3$ (Theorem 3.4) by constructing balanced separators. Via a result of Dvořák and Norin [DN19], this implies that these graphs have treewidth linear in k. However, Proposition 8.5 of Wood and Telle [WT07] implies that every outer k-planar graph has treewidth at most $3k + 11$, i.e., a better bound on the treewidth than applying the result of Dvořák and Norin to our separators. The treewidth $3k + 11$ bound also implies a separation number of $3k + 12$, but our bound is lower. Our separators also allow outer k-planarity testing in quasi-polynomial time – see Theorem 3.5.

Theorem 3.4. *Each outer k-planar graph has separation number at most $2k + 3$.*

Proof. Consider an outer k-planar drawing. If the graph has an edge that cuts between $\frac{n}{3}$ and $\frac{2n}{3}$ vertices to one side, we can use this edge to obtain a balanced separator of size at most $k + 2$, i.e., by choosing the endpoints of this edge and a vertex cover of the edges crossing it. So, suppose no such edge exists. Consider a pair of vertices a and b such that the line $\overline{ab}$ divides the drawing into a left side S_ℓ and a right side S_r having an almost equal number of vertices $(\left| |S_\ell| - |S_r| \right| \leq 1)$. If the edges which cross $\overline{ab}$ also mutually cross each other, there can be at most k of them. Thus, we again have a balanced separator of size at most $k + 2$. So, it remains to consider the case when we have some pair of edges that both cross $\overline{ab}$, but do not cross each other. We call such a pair of edges *parallel*. We now pick a specific pair of parallel edges:

Starting from b, let b_ℓ be the first vertex along the boundary in clockwise direction such that there is an edge (b_ℓ, b'_ℓ) that crosses the line $\overline{ab}$. Symmetrically, starting from a, let a_r be the first vertex along the boundary in clockwise direction such that there is an edge (a_r, a'_r) that crosses the line $\overline{ab}$; see Figure 3.2 (a). Note that the edges (a_r, a'_r) and (b_ℓ, b'_ℓ) are either identical or parallel. In the former case, we see that all other edges crossing line $\overline{ab}$ must also cross the edge $(a_r, a'_r) = (b_\ell, b'_\ell)$, and as such there are again at most k edges crossing $\overline{ab}$. In the latter case, there are two subcases to be considered. For two vertices u and v, let $[u, v]$ be the set of vertices that starts with u and, going clockwise, ends with v (with respect to the embedding). Let $]u, v[$ be the set of vertices between u and v, thus $]u, v[= [u, v] \smallsetminus \{u, v\}$.

- **Case 1:** The edge (b_ℓ, b'_ℓ) cuts $\frac{n}{3}$ vertices to the top; see Figure 3.2 (b). In this case, either $[b'_\ell, b]$ or $[b, b_\ell]$ has to have between $\frac{n}{3}$ and $\frac{n}{2}$ vertices. We claim that neither the line $\overline{bb_\ell}$ nor the line $\overline{bb'_\ell}$ can be crossed more than k times. Namely, each edge that crosses $\overline{bb_l}$ also crosses edge (b_ℓ, b'_ℓ). Similarly, each edge that crosses $\overline{bb'_\ell}$ also crosses (b_ℓ, b'_ℓ). Thus, we have a separator of size at most $k + 2$, regardless of whether we choose $\overline{bb_\ell}$ or $\overline{bb'_\ell}$ to separate the graph. As we observed above, one of them is balanced. The case where edge (a_r, a'_r) cuts at most $\frac{n}{3}$ vertices to the bottom is symmetric.

- **Case 2:** The pair of edges (b_ℓ, b'_ℓ) and (a_r, a'_r) each cut at most $\frac{n}{3}$ vertices to the bottom and top, respectively, and don't cross each other. We show that in this case, we can always find a *close* pair of parallel edges, that is, a pair of edges, both cutting $\frac{n}{3}$ vertices to the top and bottom, respectively, and with no edges between them parallel to either. If there is an edge e between (b_ℓ, b'_ℓ) and (a_r, a'_r), we form a new pair by using e to replace either (a_r, a'_r) or (b_ℓ, b'_ℓ), depending on whether e cuts at most $\frac{n}{3}$ vertices to the bottom or to the top, respectively. By repeating this procedure, we always find a close pair. Hence, we can assume that (u_b, v_b) and (u_a, v_a) actually form a close pair. Let $\alpha = |]v_a, u_a[|$, $\beta = |]v_b, u_b[|$, $\gamma = |]u_a, v_b[|$, and $\delta = |]u_b, v_a[|$ be the numbers of vertices in the four quadrants defined by the endpoints; see Figure 3.2 (c).

  Suppose that the edge pair shares an endpoint – w.l.o.g. assume $v_a = u_b$. We can now use both edges (v_a, v_b) and (u_a, v_a) (together with any edges crossing them) to obtain a separator of size at most $2k+3$. The separator is balanced since $\alpha+\beta \le \frac{2n}{3}$ and $\gamma + \delta \le \frac{2n}{3}$.

  So, instead assume that the endpoints u_a, v_a, u_b, v_b are all distinct vertices. Note that we have $\gamma, \delta \le \frac{n}{2}$ since each side of $\overline{ab}$ has at most $\frac{n}{2}$ vertices. We separate the

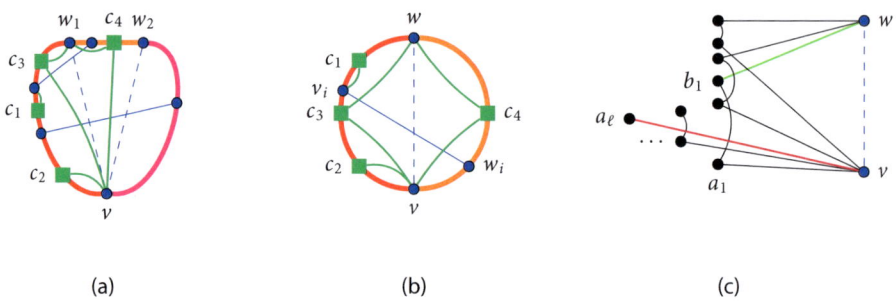

Figure 3.3: Shapes of separators, special separator S in blue, regions in different colors (red, orange, and pink), components connected to blue vertices in green: (a) closest-parallels case; (b) single-edge case; (c) special case for single-edge separators.

graph along the line $\overline{u_b u_a}$ as follows: All the edges that cross this line must also cross (u_b, v_b) or (v_a, u_a). Therefore, we obtain a separator of size at most $2k + 2$.

To see that the separator is balanced, we consider two cases. If $\delta \geq \frac{n}{3}$ (or $\gamma \geq \frac{n}{3}$), then $\alpha + \beta + \gamma \leq \frac{2n}{3}$ (or $\alpha + \beta + \delta \leq \frac{2n}{3}$). Otherwise we have $\delta < \frac{n}{3}$ and $\gamma < \frac{n}{3}$. In this case $\delta + \alpha \leq \frac{2n}{3}$ and $\gamma + \beta \leq \frac{2n}{3}$. In both cases the separator is balanced.

$\square$

Using the existence of balanced separators in outer k-planar graphs, we get the following result.

Theorem 3.5. *The outer k-planarity of a graph with n vertices can be tested in $O(2^{\text{polylog } n})$ time when k is fixed.*

Proof. Our approach is to leverage the structure of the balanced separators as described in the proof of Theorem 3.4. Namely, we enumerate the sets which could correspond to such a separator, pick an appropriate outer k-planar drawing of these vertices and their edges, partition the components arising from this separator into *regions*, and recursively test the outer k-planarity of the regions.

To obtain quasi-polynomial runtime, we need to limit the number of components on which we branch. We do this by grouping them into regions defined by special edges of the separators.

By the proof of Theorem 3.4, if our input graph has an outer k-planar drawing, there must be a separator which has one of the two shapes depicted in Figure 3.3 (a) and (b). Here we are not only interested in the up to $2k + 3$ vertices of the balanced separator, but actually the set S of up to $4k + 3$ vertices one obtains by taking both endpoints of the edges used to find the separator. Note that this larger set S is also a balanced separator. We use a brute force approach to find the right set: Enumerating all sets of vertices of size up to $4k + 3$, we then check whether it can be drawn similar to one of the two shapes from Figure 3.3. By fixing S, we pick a subgraph G_S induced by S with $O(k)$ vertices. Graph G_S can have at most a function of k different outer k-planar drawings. Thus, we branch on all such drawings of G_S.

We now consider the two different shapes separately. In the first case, S contains three special vertices v, w_1 and w_2; in the second case S contains two special vertices v and w. In both cases, the special vertices will be called *boundary* vertices and all other vertices in S will be called *regional* vertices. By fixing the drawing of G_S in the current branch, the regional vertices are partitioned into regions by the boundary vertices. Using the structure of the separator guaranteed by the proof of Theorem 3.4, we get that no component of $G \setminus S$ can be adjacent to regional vertices which live in different regions with respect to the boundary vertices.

We first discuss the case of using G_S as depicted in Figure 3.3 (a). We start by picking the three special vertices v, w_1 and w_2 from S to behave as shown in Figure 3.3 (a). The following arguments regarding this shape of separator are symmetric with respect to the pair of opposing regions.

If there is a component connected to regional vertices of different regions, we reject this configuration. From the proof of Theorem 3.4, no component can be adjacent to all three boundary vertices: this would either contradict the closeness of the parallel edges or imply an edge connecting distinct regions. Each component is of one of four different types, depending on how it is connected to regional and/or boundary vertices; for the regions neighboring w_1 they are shown as c_1, c_2, c_3, and c_4 in Figure 3.3 (a).

- Components of type c_1 are connected to (possibly many) regional vertices of the same region and may be connected to boundary vertices as well. In any valid drawing, they have to be placed in the same region as their regional vertices.

- Components of type c_2 are not connected to any regional vertices and only connected to one of the three boundary vertices. Hence, they cannot interfere with other parts of the drawing – we can arbitrarily assign them to a region adjacent to their boundary vertex.

- For components that are connected to two boundary vertices – say v and w_1 – it seems to be possible for them to be placed left or right of the edge connecting v and w_1, e.g., as c_3 or c_4. The latter option c_4 is not valid: The separator was created by two close parallel edges as argued in the proof of Theorem 3.4, a contradiction.

From the above discussion, we see that from a fixed configuration – a set S, a drawing of G_S, and triple of boundary vertices – if the drawing of G_S has the shape depicted in Figure 3.3 (a), we can either reject the current configuration for having bad components, or we obtain a well-defined placement into the regions defined by the boundary vertices: For components of type c_2, it suffices to recursively produce a drawing of that component together with its boundary vertex to be placed next to that boundary vertex. The other components can be partitioned into the regions and we recurse on the regions individually. This covers all cases for this separator shape.

The shape of the separator for the second case is shown in Figure 3.3 (b). We have two boundary vertices v and w and thus only two regions. This allows for the two component types c_1 and c_2 from the first case. They are also handled as described above. In addition, we have components connected to both v and w but no regional vertices. This allows for two different placement options c_3 or c_4 – left or right of the line $\overline{vw}$. If there is an edge (v_i, w_i) that is part of the separator but different from vw, there cannot be more than a total of k such components; see Figure 3.3 (b). In any drawing, there will be edges connecting each component to v and w, and at least one of these edges has to cross (v_i, w_i). Since this edge can be crossed at most k times, this gives an upper bound on the number c_3 and c_4-type components. Thus, we now enumerate all different placements of these components as either type c_3 or type c_4 and recurse accordingly.

In the case that there is no edge (v_i, w_i), the separator is exactly the pair v, w. This eliminates the possibility of type c_1 components and the components of type c_2 are handled as before. We argue that in this case, any a valid drawing can have at most a function of k different components of type c_3 or c_4. Consider the components of type c_3, the components of type c_4 can be counted similarly. Given a valid drawing of a type c_3 component, consider the highest – clockwise last – vertex of this drawing connected to v and the lowest – clockwise first – vertex connected to w. These two vertices define a subinterval of the left region. Considering two such intervals, they can relate in one of three ways: They overlap, they are disjoint, or one is contained in the other. We group components with either overlapping or disjoint intervals into *layers*. This is shown in Figure 3.3 (c): For simplicity, we only draw the highest and lowest connected vertices for every component and we contract every component into a single edge representing connectivity.

Let (a_1, b_1) be the bottom-most component of type c_3 – vertex a_1 is the lowest of all lower vertices. We define the first layer to be all components overlapping or disjoint of (a_1, b_1). Now consider the (green) edge (b_1, w) (see Figure 3.3 (c)): The total number of components disjoint from (a_1, b_1) in the first layer is bounded by k since for every component, at least one of its edges connecting it to v must cross (b_1, w). Intervals that overlap (a_1, b_1) must have an edge connecting the vertex inside (a_1, b_1) to either v or w. This edge then must cross either (a_1, v) or (b_1, w). This implies that there can only be $O(k)$ components in the first layer.

New layers are defined by considering components whose intervals are fully contained in the first interval of the previous layer, starting with the second layer completely being contained inside (a_1, b_1). Notice that intervals fully contained in other intervals disjoint from (a_1, b_1) are also in the first layer and accounted for accordingly. All deeper layers are nested inside (a_1, b_1). To limit the total depth, let a_ℓ be the lowest vertex of the first component of the deepest layer. Consider the (purple) edge (v, a_ℓ) – it is crossed by some edge of every layer above it. As any edge can only have k crossings, there can only be $O(k)$ different levels in total. This leaves us with a total of at most $O(k^2)$ components per region and again we can enumerate their placements and recurse accordingly.

The above algorithm provides the following recurrence regarding its runtime. Denote by $T(n)$ the runtime of our algorithm which is generously upper bounded by the following expression. Let $f(s)$ be the number of different outer k-planar drawings of a graph with s vertices.

$$T(n) \leq \begin{cases} n^{O(k)} \cdot f(4k+3) \cdot n^3 \cdot n \cdot T(\frac{2n}{3}) & \text{for } n > 5k \\ f(n) & \text{otherwise.} \end{cases}$$

Thus, the algorithm runs in quasi-polynomial time, i.e., $2^{\text{poly}(\log n)}$. $\qquad\square$

3.3 Outer k-Quasi-Planar Graphs

In this section we consider outer k-quasi-planar graphs. We first describe some classes of graphs which are (not) outer 3-quasi-planar. We then discuss edge-maximal outer k-quasi-planar drawings.

3.3.1 Comparability to Planar Graphs

A *sub-Hamiltonian* graph is a planar graph to which edges can be added such that the resulting augmented graph is still planar and contains a Hamiltonian cycle. Note that all sub-Hamiltonian planar graphs are outer 3-quasi-planar: The (completed) Hamiltonian cycle divides the edges into two sets of mutually non-crossing edges – one outside and one inside the cycle. Flipping the outside edges to the inside, we get sets of pairwise crossing edges of size at most 2. In this embedding, the edges completing the Hamiltonian cycle can be added without crossings (they bound the outer face). One can also see which complete and bipartite complete graphs are outer 3-quasi-planar.

Proposition 3.1. *The following graphs are outer 3-quasi-planar:*

(a) *$K_{4,4}$ and K_5,*

(b) *planar 3-tree with three complete levels, and*

(c) *square-grids of any size.*

Proof. It is easy to verify (a) by constructing a valid drawing, such as those in Figure 3.4.

(b) was experimentally verified by using MINISAT [SE05] to check a Boolean expression for satisfiability. The details on the SAT formulation can be found in Section 3.6.1.

(c) follows from an old result by Chung, Leighton, and Rosenberg [CLR87]: Square-grids are sub-Hamiltonian. $\qquad\square$

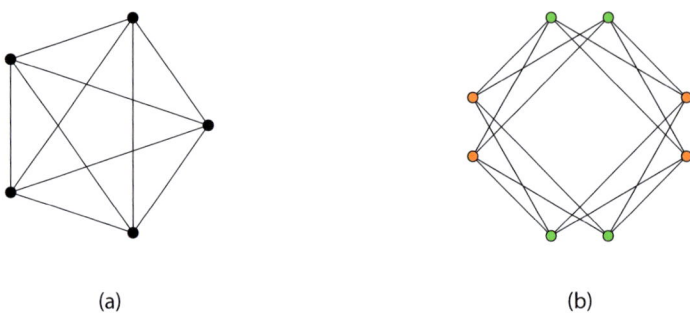

(a)   (b)

Figure 3.4: Illustrations for Proposition 3.1: An outer 3-quasi-planar embedding of (a) K_5 and (b) $K_{4,4}$.

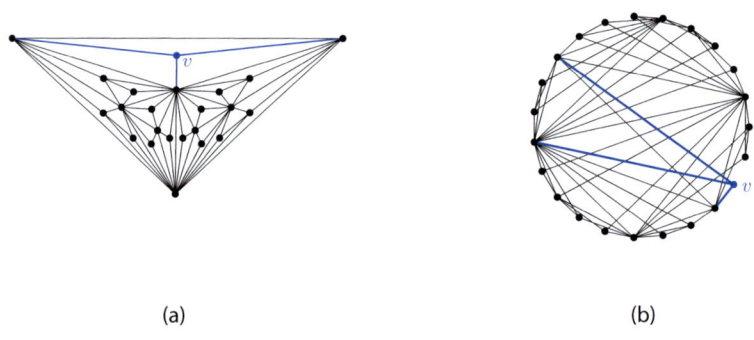

(a)   (b)

Figure 3.5: A vertex-minimal 23-vertex planar 3-tree which is not outer quasi-planar: (a) planar drawing; (b) deleting the blue vertex makes the drawing outer quasi-planar.

In addition, we state in the proposition below that certain complete and complete bipartite graphs which are not outer 3-quasi-planar. The family of *planar 3-trees* is the set of graphs obtainable by the following construction: Take a planar drawing of the complete graph K_4 and repeat the *stacking* operation as often as desired: Pick some inner face f of the drawing, add a vertex v at f's center, and connect v to all vertices incident to f (subdividing f into three new inner faces). By convention K_4 is the planar 3-tree with one *complete level*; performing the stacking operation on all original inner faces of a planar 3-tree with i complete levels, a planar 3-tree with $i + 1$ complete levels is created.

Proposition 3.2. *The following graphs are not outer 3-quasi-planar:*

(a) *complete bipartite graphs $K_{p,q}$ with $p \geq 3$, $q \geq 5$,*

(b) *complete graphs K_n with $n \geq 6$, and*

(c) *planar 3-tree with four complete levels.*

Furthermore, not all planar graphs have an outer quasi-planar drawing. Consider the vertex-minimal planar 3-tree on 23 vertices shown in Figure 3.5 (a); using MINISAT to check for satisfiability of the corresponding Boolean expression we verified that it is not outer quasi-planar. An almost outer-quasi planar drawing of this graph can be seen in Figure 3.5 (b). It was constructed by removing the blue vertex, drawing the remaining graph in an outer quasi-planar way, and then reinserting the missing vertex. A description of the formula used to verify this result is found in Section 3.6.1.

Together, Propositions 3.1 and 3.2 immediately yield the following.

Theorem 3.6. *Planar graphs and outer 3-quasi-planar graphs are incomparable under containment.*

3.3.2 Maximal Outer *k*-Quasi-Planar Graphs

If adding any edge to a drawing of an outer k-quasi-planar graph destroys the outer k-quasi-planarity of that drawing, it is called *maximal*. We call an outer k-quasi-planar graph maximal if it has a maximal outer k-quasi-planar drawing. Recall that Capoyleas and Pach [CP92] showed the following upper bound on the edge density of outer k-quasi-planar graphs on n vertices:

$$|E| \leq 2(k-1)n - \binom{2k-1}{2}.$$

We prove that each maximal outer k-quasi-planar graph meets this bound. We include our independently discovered proof even though two other proofs of this result can be found in the literature – by Dress, Koolen, and Moulton [DKM02] and also by Nakamigawa [Nak00], we thank David Wood for pointing us to them – the main result of both papers prove a slightly stronger theorem: For a drawing $G = (V, E)$, an *edge flip* produces a new drawing G^* by replacing an edge $e \in E$ with a new edge $e^* \in \binom{n}{2} \setminus E$. Both works show that, for every two edge-maximal outer k-quasi-planar drawings $G = (V, E)$ and $G' = (V, E')$, there is a sequence of edge flips producing drawings $G = G_1, G_2, \ldots, G_t = G'$ such that each G_i is a maximal k-quasi-planar drawing. Together with the tight example of Capoyleas and Pach, this implies the next theorem. The proof we present here directly employs an inductive argument, building on the ideas of Capoyleas and Pach. The argument itself and the structural insights leading up to it are of independent interest.

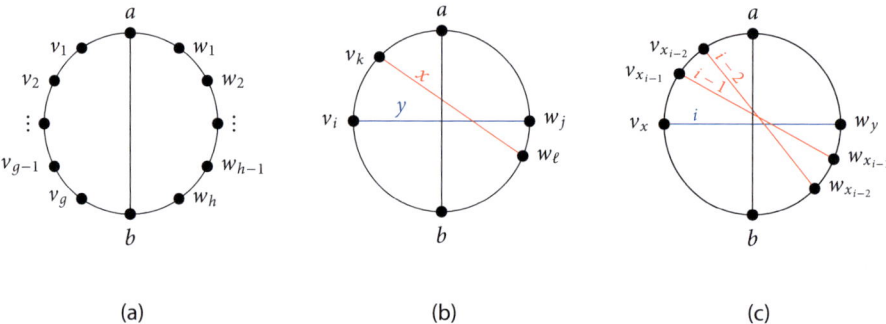

Figure 3.6: Long edges in outer k-quasi-planar graphs: (a) Labeling scheme for left and right side vertices according to the long edge $\overline{(a,b)}$. (b) Illustration of property (P1): $(v_k w_\ell)$ is crossing (v_i, w_j) from above. (c) Property (P2): Blue edge in level i and the first two edges of a possible certificate.

Theorem 3.7. *Each maximal outer k-quasi-planar drawing $G = (V, E)$ has:*

$$|E| = \begin{cases} \binom{|V|}{2} & \text{if } |V| \leq 2k - 1, \\ 2(k-1)|V| - \binom{2k-1}{2} & \text{if } |V| \geq 2k - 1. \end{cases}$$

For an outer k-quasi-planar drawing of graph G we call an edge $\overline{(a,b)}$ a *long edge*, if a and b are separated along the outer face of G by at least $k - 1$ vertices on both sides. In the following depictions, long edges will always be drawn vertically with a on top, dividing the graph into the two regions left and right of $\overline{(a,b)}$. All edges that intersect the long edge are called *crossing edges*. All vertices incident to crossing edges will be called *crossing vertices* and for illustration we will label them as follows: In the left region, vertices will be labeled $v_1, \ldots, v_g$ counterclockwise from a on – from top to bottom –, and in the right region, vertices will be labeled $w_1, \ldots, w_h$, and by definition, we have $g, h \geq k - 1$; see Figure 3.6 (a).

To prove maximality of the considered graph, we count the number of crossing edges in an inductive argument. We construct $(k - 2)$ *hierarchical levels* – subsets of the crossing edges of G that form maximal crossing-free connected subgraphs of G. We then define a replacement-operation that uses these levels to split the original graph into two subgraphs, each with fewer vertices.

We use Algorithm 3.1 to greedily build the hierarchical levels. The correctness of Algorithm 3.1 is shown in the following lemma.

Lemma 3.8. *For a given edge-maximal outer k-quasi-planar graph G, Algorithm 3.1 generates $k - 2$ hierarchical levels.*

Proof. We need to argue two things: The algorithm creates $k - 2$ levels that cover all crossing edges with respect to $\overline{(a,b)}$, and that every level is connected. We order the

---

Algorithm 3.1: BUILDLEVELS(Outer k-quasi-planar Graph G, long edge $\overline{(a,b)}$)

Vertices $\{v_1, \ldots, v_g\} \leftarrow$ vertices left of $\overline{(a,b)}$
Set of Sets $\mathcal{S} = \{S_1, \ldots, S_{k-2}\} \leftarrow$ empty level sets
for $i \leftarrow 1$ **to** $k-2$ **do**
    for $j \leftarrow 1$ **to** g **do**
        $S' \leftarrow$ edges of v_j not crossing edges in S_i
        $S_i \leftarrow S_i \cup S'$

return $\mathcal{S}$

---

levels by order of construction: level y is after level x or $x \prec y$, if y is constructed after x. To do so, we first state two important properties:

(P1) If an edge (v_i, w_j) of level y is crossed by an edge (v_k, w_ℓ) of level x with $x \prec y$, then (v_k, w_ℓ) must cross (v_i, w_j) *from above*: $i > k$ and $j < \ell$; see Figure 3.6 (b).

(P2) For any edge e of level i, there is a set of edges $\mathcal{E} = \{e_1, \ldots, e_{i-1}\}$ – one from each previous level – such that $\mathcal{E} \cup e$ is a set of i pairwise crossing edges; see Figure 3.6 (c).

Property (P1) follows from the construction of level x: If there where no edge of level x crossing edge (v_i, w_j), then (v_i, w_j) would also belong to level x.

Property (P2) holds by induction: For the first level there is no previous level. Edges of level two are crossed by edges of the first level due to (P1). Any edge (v_x, w_y) of level i must be crossed by some edge $(v_{x_{i-1}}, w_{y_{i-1}})$ of level $i-1$. Inductively we know that e_{i-1} is crossed by an edge of every previous level. Together, they form a chain of pairwise crossings from above, and we get the following patterns on the indices of these edges:

$$x > x_{i-1} > x_{i-2} > \cdots > x_1 \text{ and } y < y_{i-1} < y_{i-2} < \cdots < y < y_1.$$

These patterns indicate that in fact all the considered edges are pairwise crossing. For a given edge e of level i, we call any set following the description in property (P2) a *certificate* for e to be in i and any edge of level $i-1$ crossing e can be extended to some certificate for e.

Let t be the last level and consider the last edge $e^\dagger$ taken from that level. Suppose the algorithm created too many levels, so $t > k-2$. By property (P2), the certificate of $e^\dagger$ belonging to t together with $e^\dagger$ and $\overline{(a,b)}$ forms a set of $t+1 \geq k$ pairwise crossing edges. This contradicts that the graph from which $e^\dagger$ is taken is outer k-quasi-planar. Hence, we never create more than $k-2$ levels.

As Algorithm 3.1 greedily takes any legal edge into the current set, each level is maximal by construction.

To argue about the connectivity of the levels we carefully consider the way they will be generated in a maximal graph. We ensure that greedily picking edges never disconnects

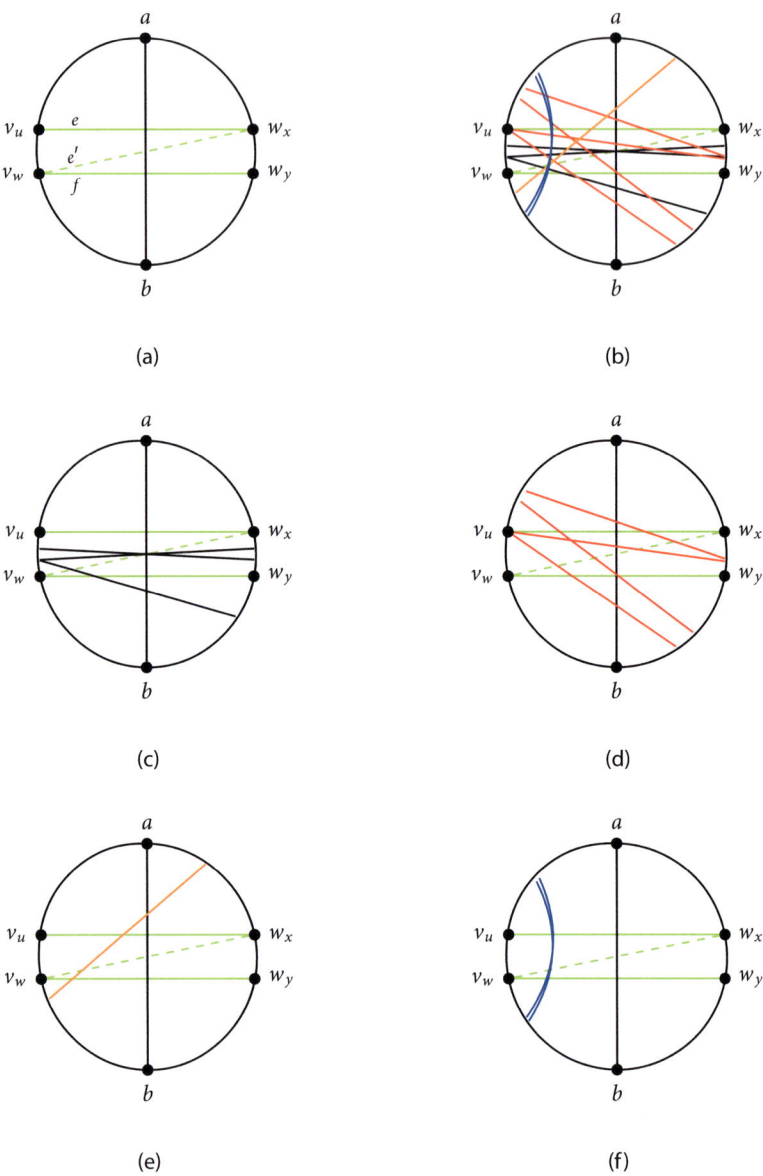

Figure 3.7: Connectivity of hierarchical level i: (a) The green level i is disconnected, as dashed edge e' is missing. (b) A full set of edges blocking e' using only locally left edges; (c) black edges start between v_u and v_w, but do not belong to the same level as e and f, (d) red edges cross e' from above and end below w_x, (e) orange edges cross e' from below, and (f) blue edges start and end on the same side.

the edge set of any level. Assume we already generated levels 1 to $i - 1$ and in level i we pick the edges e and f but not e', as shown in Figure 3.7 (a).

Assume that edge e' exists but is not put into level i, ending up with that level being disconnected. Consider the relative positions of the endpoints of e' in the left and right regions with respect to the other edges of level i. As e' is going upward it cannot cross other edges of level i from above due to Property (P1). Thus, if e' is present in our graph, it would belong to level i and i would be connected, a contradiction.

The other option to have level i be disconnected is that edge e' is missing from the graph. As the graph is supposed to be maximal, there must be a *prevention* set – a set of $k - 1$ pairwise crossing edges that prevent its existence. Due to the edge $\overline{(a,b)}$, the prevention set can contain at most $k - 2$ crossing edges. By existence of the edges e and f we know that the prevention set also cannot consist of edges only locally on one side: Any prevention set locally left would also prevent the existence of f by covering v_w; a symmetric argument can be made against a locally right set together with edge e and vertex w_x.

In this context, $\overline{(a,b)}$ is considered to be both locally left and locally right. Hence, any prevention set would have to include crossing edges as well, implying two possible options: crossing edges together with locally left or right edges – see Figure 3.7 (b). For illustration we use colors to describe the different types of edges of the prevention set depicted in Figure 3.7 (b) differently: black edges (Figure 3.7 (c)) start between v_u and v_w, appearing between e and f but not belonging to the same level as e; red edges (Figure 3.7 (d)) cross e' from above ending strictly below w_x; orange edges Figure 3.7 (e) cross e' from below; blue edges (Figure 3.7 (f)) are locally on one side crossing all edges of other colors. By verifying the following claim, we finish the proof.

Claim: A prevention set $\mathcal{P}$ for e' can be transformed into a prevention set for f (or symmetrically e) – contradicting the existence of that set.

To transform a prevention set $\mathcal{P}$ for e' into a prevention set $\mathcal{P}'$ for f, we proceed as follows: We keep all the orange and blue edges of $\mathcal{P}$, as they already pairwise cross and also cross f. If we replace any of the remaining edges, we need to make sure that the new edges also cross the blue and orange ones in $\mathcal{P}'$.

Any other edges in $\mathcal{P}$ belong to previous levels because of the following:

- By (P1), all red edges cross e' from above.

- For the black edges, we have two different cases: Each edge either crosses f or lives completely between e and f. Again by (P1), edges crossing f do so from above. Edges that live between e and f cannot be crossed by edges of level i, so they must belong to a level before level i.

We now consider the edges of $\mathcal{P}$ in the order of the level they belong to starting at $i - 1$. For any such level, there is an edge crossing f whose left vertex is between the two endpoints of every blue edge of $\mathcal{P}$. To see that this is true, assume that there is some edge e_k of level $k < i$ crossing e' but not f. Since f is not in level k, there must be an edge $e^\times$ crossing it and its right vertex must be below the right vertex of e_k. This leaves

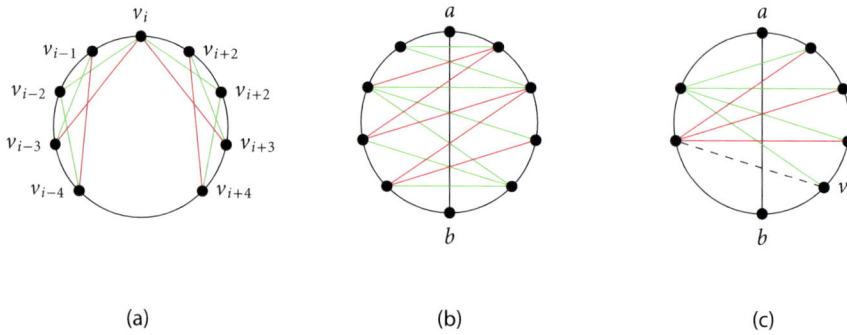

Figure 3.8: Using hierarchical levels to perform the split operation: (a) The frame of a maximal outer k-quasi-planar graph; black edges are present for any k, green edges for $k \geq 3$, purple edges for $k \geq 4$). (b) An outer 4-quasi-planar graph with 2 hierarchical levels shown in green and purple respective; (c) the resulting subgraph G_1, the presumably missing edge of v (drawn dashed) is actually a frame edge.

to options for the left vertex of $e^\times$: it can either also be the left vertex of e_k or some other vertex below it but still above v_w; in both cases, this left vertex will be covered by all blue edges and we can replace e_k by $e^\times$. In the worst case, performing this replacement stops all remaining edges of the certificate for e' from crossing $e^\times$ because it is below them. As $e^\times$ is in level k, there must be a prevention set for that. Notice that every edge of this certificate must be below its corresponding counterpart of the certificate for e' and thus still all starting vertices are covered by blue edges. Hence then we also replace all edges of layers before k by edges of the certificate for $e^\times$.

This conclude that a prevention set for e' using locally left edges can be transformed into a prevention set for f. A similar argument can be made using locally right edges by again taking a prevention set for e', transforming the certificate for e' into one for e and observing that the endpoints of of the edges of the certificate are again still covered by the locally right edges.

We have established that edge e' must exist and has to be placed into the same layer as e and f, connecting the two components. □

Considering the construction of Pach and Capoyleas [CP92], the proof of Claim 4 yields the following:

Remark. Let G be an outer k-quasi-planar graph with n vertices and let the vertices be labeled $v_1, v_2, \ldots, v_n, v_{n+1} = v_1$ according to their cyclic order along the outer face. Every vertex v_i can be adjacent to v_ℓ with $\ell \in [i - (k-1), i + (k-1)]$. Hence, these *frame* edges are present in any maximal outer k-planar graph; see Figure 3.8 (a).

Using the hierarchical levels described above, we give a *split* operation, which is used to split the graph into two smaller parts. Let G be a maximal outer k-quasi-planar graph with a long edge $\overline{(a,b)}$ and hierarchical levels created by Algorithm 3.1. Let L_i and R_i be the vertices of G incident to the crossing edges of level i in the regions left and right of $\overline{(a,b)}$, respectively. Splitting G into two subgraphs G_1 and G_2 is done as follows: To obtain G_1, for every level i from 1 to $(k-2)$, replace the vertices of L_i by a single *level*-vertex v_i and connect that vertex to all vertices in R_i, see Figure 3.8 (b) and (c). Finally, add to G_1 all missing frame edges to make it maximal. To obtain G_2, proceed symmetrically, exchanging the roles of left and right.

Lemma 3.9. *After applying a split operation to a maximal outer k-quasi-planar graph G, the following relations among G and its two subgraphs G_1 and G_2 hold:*

(i) $|V(G)| = |V(G_1)| + |V(G_2)| - 2k + 2$ *and*

(ii) $|E(G)| = |E(G_1)| + |E(G_2)| - (|E_1'| + |E_2'|) + |E'| - 1$,

where E' is the set of edges of G crossing $\overline{(a,b)}$ and E_1', E_2' are the sets of crossing edges added to G_1, G_2 by the split operation.

Proof. We establish the two equations (i) and (ii) individually.

(i): Graphs G_1 and G_2 are obtained by only modifying vertices on the right or left side of G respectively, leaving the other side unmodified and $\overline{(a,b)}$ present in both. The modification adds $k - 2$ level-vertices to each graphs, so subtracting these vertices and one copy of the vertices a and b, yields:

$$|V(G)| = |V(G_1)| - (k-2) + |V(G_2)| - (k-2) - 2$$
$$= |V(G_1)| + |V(G_2)| - 2k + 2.$$

(ii): We count the edges added to both G_1 and G_2 and compare them to the number of edges removed by splitting G. To do so, we consider the structure of our hierarchical levels and the respective left and right vertices L_i and R_i. Every level is a caterpillar – a set of connected and non-crossing edges –, so there are exactly $|L_i| + |R_i| - 1$ edges in level i. Hence, the total number of crossing edges over all levels is

$$|E'| = \sum_{i=1}^{k-2}(|L_i| + |R_i| - 1).$$

The sets of edges added to G_1 and G_2 each consist of two different types of edges: The first type are the edges incident to the level-vertices, the other type are the missing frame edges added to ensure that G_1 and G_2 are maximal. The total number of edges of the first type can be expressed by summing up the sizes of all sets L_i and R_i with $1 \leq i \leq k - 2$ as follows:

$$\sum_{i=1}^{k-2} |R_i| + |L_i|.$$

The subgraphs induced by vertices and edges of the unmodified sides of G_1 and G_2 remain maximal by maximality of G. But considering the last vertex of some level, in some cases it seems possible to add an additional edge incident to that vertex to the drawing – for instance, see vertex v from Figure 3.8 (c). To count the presumably missing edges, recall how we generated the hierarchical levels. In level i we take all remaining edges of right-side vertex $w_{h-(i-1)}$. For the first level, we take all edges of the last right-side vertex, for the second level all edges of the second-to-last vertex, and so on. The total number of edges missing this way in both subgraphs together is

$$2 \cdot \sum_{i=1}^{k-2} (i-1).$$

The distance of each level vertex to the lowest vertex on the other side in the cyclic order – counted clockwise or counterclockwise, depending on in which region the level vertex was placed – is by construction bounded by $k-1$. Hence, any edge missing this way is actually a frame edge that can safely be added to the subgraph to make it maximal.

We added $k-2$ level vertices and the maximum number of frame edges incident to them. As G is a k-quasi-planar graph, the vertices connecting these frame edges together with a and b form a clique on $k - 2 + 2$ vertices. The number of edges in this clique, and hence the exact total amount of edges we add to G_1 and G_2 this way (without counting $\overline{(a,b)}$) is

$$2\left(\binom{k}{2} - 1\right).$$

Putting everything together, the total number of new edges added to G_1 and G_2 is

$$|E_1'| + |E_2'| = 2\left(\binom{k}{2} - 1\right) + \sum_{i=1}^{k-2} R_i + L_i + 2(i-1).$$

Notice that we did not account for $\overline{(a,b)}$ in any of the subgraphs yet, so we subtract one copy of it. The remaining parts of G_1 and G_2 are the unmodified sides of G and one copy of $\overline{(a,b)}$. Simplifying the above equation completes the proof. □

To complete our inductive argument, we need to do two things: We need to prove the existence of a long edge in any maximal outer k-quasi-planar graph and we need to consider the base cases – those maximal outer k-quasi-planar graphs with the minimum number of vertices.

Lemma 3.10. *Any maximal outer k-quasi-planar graph $G = (V, E)$ either*

(i) *is a clique of size $|V| \leq 2k - 1$, or*

(ii) *has a long edge.*

Proof. We consider each case individually, depending on number of vertices in the graph.

For case (i), when $|V| \le 2k - 1$, the graph G has to have all possible frame edges by maximality. For any vertex v these edges connect it to the closest $k - 1$ other vertices left and right of it. As any v together with its frame edge neighbors covers all vertices of G, the graph itself is the complete graph K_{2k-1}.

In case (ii) we have $|V| > 2k - 1$: Doing a simple counting argument on the number of neighbors induced by frame edges, we get that for every vertex v_i ($i \in 1, \dots, n$), there is at least one other vertex w_i that it is not connected to via a frame edge. Consider the pair v_i, w_i for some i. If it were connected by an edge, this would be the long edge we are looking for. So suppose the edge (v_i, w_i) is missing. As we choose G to be maximal, there must be some set S of edges preventing the existence. As edges of S cannot be part of the frame, they must span more than $k - 1$ vertices on both sides. Hence, S must contain at least on long edge. $\square$

Finally, we combine the results of the lemmata presented above to complete the proof of Theorem 3.7.

Proof of Theorem 3.7. By Lemma 3.10, the graph G we consider in each step either (i) is a clique or (ii) we always find a long edge to split by. By picking a long edge in G, dividing it into two regions, building the hierarchical levels with respect to these regions and performing the split operation as described above, we get two subgraphs G_1 and G_2 of smaller size. By Lemma 3.9 we get the relationships on vertex- and edge-count between G and both subgraphs. We recursively repeat the splitting on these subgraphs until we encounter cliques. We then know that the number of edges matches the bound on total edge number.

Given a maximal outer k-quasi-planar graph G, we can recursively split it into pieces that individually retain maximality and outer k-quasi-planarity. Considering the relationship among the edge sets of Lemma 3.9 (ii) and the maximality of the subgraphs, we get the following equation:

$$
\begin{aligned}
|E(G)| &= |E(G_1)| + |E(G_2)| - (|E_1'| + |E_2'|) + |E'| - 1 \\
&= |E(G_1)| + |E(G_2)| - 2k^2 + 5k - 3 \\
&= 2(k - 1)(n_1 + n_2) - 2\binom{2k-1}{2} - 2k^2 + 5k - 3.
\end{aligned}
$$

From the maximality of G and using Lemma 3.9 (i), we now have:

$$
\begin{aligned}
2(k - 1)(-2k + 2) &= -\binom{2k-1}{2} - 2k^2 + 5k - 3 \\
-4k^2 + 8k - 4 &= -4k^2 + 8k - 4.
\end{aligned}
$$

The equation balances, completing the proof. $\square$

3.4   Testing for Full Convex Drawings via MSO₂

Hong and Nagamochi [HN16] were the first to introduce the class of *full outer k-planar graphs*. Graphs in this class have a convex drawing which is outer k-planar and additionally there is no crossing on the outer boundary of the drawing – every corner of the (not necessarily simple) polygon prescribing the outer face is a vertex of the graph. Hong and Nagamochi gave a linear-time recognition algorithm for full outer 2-planar graphs. They state that a graph G is (full) outer-2-planar, if and only if its biconnected components are (full) outer-2-planar and that the outer boundary of a full outer-2-planar embedding of a biconnected graph G is a Hamiltonian cycle of G. In Theorem 3.11, we observe that this property also carries over to general outer k-planar and outer k-quasi-planar graphs. Therefore, we define the classes of *closed outer k-planar* and *closed outer k-quasi-planar* graphs, where closed means that there is an appropriate convex drawing where the circular order forms a Hamiltonian cycle.

In the following, we first give a basic introduction to Monadic Second-Order Logic (MSO₂) and Courcelles' Theorem [Cou90, CE12], then use MSO₂ to express crossing patterns of closed k-planar and k-quasi-planar graphs. This will result in a linear-time algorithm to test closed outer k-planarity for each fixed k. The following Theorem 3.11 will then conclude this section, translating the algorithm from closed to full outer k-planar graphs.

Theorem 3.11. *To test a given graph for full outer k-planarity or outer k-quasi-planarity it suffices to test its biconnected components for closed outer k-planarity or outer k-quasi-planarity respectively.*

Proof. Let G be a graph. Clearly, if a subgraph is not closed outer k-planar, then neither is the whole graph. It remains to show that when all biconnected components are all closed outer k-planar, then the whole graph is full outer k-planar. It is well-known that the set of cut vertices of G can be obtained in linear time. Splitting G at the cut vertices, we obtain biconnected closed components if and only if G was full itself: When one component only has drawings showing some crossing on the outside, this carries over to a drawing of the full graph – since the biconnected components of a graph form a tree, attaching the other components cannot close a cycle that covers a side of a component.

Every biconnected component is considered individually and checked in linear time. Each non-cut vertex belongs to exactly one component and thus is handled only once. The total effort spent on considering cut vertices is bounded by the number of biconnected components. Hence, a simple charging argument on the vertices concludes runtime analysis.

The MSO₂ formulas stated below in Section 3.4.2 guarantee that the Hamiltonian cycle – if present – is placed on the outer boundary of the drawing of each component. Putting together the individual drawings of the components can be done crossing free be reidentifying the cut vertices. □

3.4.1 Introduction to Monadic Second-Order Logic

Monadic Second-Order Logic (MSO$_2$) – a subset of *second-order logic* – can be used to express certain graph properties. Formulas in MSO$_2$ can be built using these primitives:

- Variables for vertices, edges, sets of vertices, and sets of edges;

- Binary relations for equality ($=$), membership in a set ($\in$), subset of a set ($\subseteq$), and edge–vertex incidence (I);

- Standard propositional logic operators: $\neg$, $\wedge$, $\vee$, $\rightarrow$, and $\leftrightarrow$;

- Standard quantifiers ($\forall$, $\exists$) which can be applied to all types of variables.

For a graph G and an MSO$_2$ formula ϕ, we use $G \vDash \phi$ to indicate that ϕ can be satisfied by G in the obvious way. Properties expressed in this logic allow us to use the powerful algorithmic result of Courcelle stated in the next section.

Any formula presented here assumes that a graph G is given and uses edges, vertices and incidences of G. To simplify notation in the following, we will always use e and f as variables for edges, F as a set of edges, u, v as vertices and U as a set of vertices (also including sub- and superscripted variants). In addition to the quantifiers above we also use a logical shorthand for the existence of exactly x elements ($\exists^{=x}$) satisfying a property, that are all pairwise unequal and that no $x+1$ such elements exist. The following formula allows us to describe connectedness of the subgraph induced by an edge set F.

$$\textsc{Connected-Edges}(F) \equiv (\forall F' \subset F)[\exists e, f \colon e \in F' \wedge f \in F \smallsetminus F'] \wedge$$

$$\Big((\exists f, e, e^* \in F \colon e \in F' \wedge f \notin F')(\exists u, v)[I(u, f) \wedge I(v, e) \wedge I(v, e^*) \wedge I(u, e^*)] \Big)$$

It states that for every proper subset F' of edge set F, we can find three edges e, f, e^* – one in F', one not in F' and e^* connecting the ends of two edges – one in F' and the other outside of F'.

The following formulas are used to describe Hamiltonicity of G. Formula $\textsc{Cycle-Set}$ implies that the edges of F form cycles, $\textsc{Cycle}$ implies maximality of the cycle and $\textsc{Span}$ forces the cycle to have an edge incident to every vertex of G.

$$\textsc{Cycle-Set}(F) \equiv (\forall e)\Big[e \in F \Rightarrow (\exists^{=2} f)\big[f \in F \wedge e \neq f \wedge (\exists v)[I(e, v) \wedge I(f, v)] \big] \Big]$$

$$\textsc{Cycle}(F) \equiv \textsc{Cycle-Set}(F) \wedge \textsc{Connected-Edges}(F)$$

$$\textsc{Span}(F) \equiv (\forall v)(\exists e)[e \in F \wedge I(e, v)]$$

$$\textsc{Hamiltonian}(F) \equiv [\textsc{Cycle}(F) \wedge \textsc{Span}(F)]$$

$\textsc{Vertex-Partition}$ implies the existence of a partition of the vertices of G into k disjoint subsets.

$$\textsc{Vertex-Partition}(U_1, \ldots, U_k) \equiv (\forall v)\left[\left(\bigvee_{i=1}^{k} v \in U_k \right) \wedge \left(\bigwedge_{i \neq j} \neg(v \in U_i \wedge v \in U_j) \right) \right]$$

3.4.2   Courcelle's Theorem and Closed Outer *k*-(Quasi-)Planarity

We now state Courcelle's Theorem and give the formulas required to express closed outer *k*-(quasi-)planarity.

Theorem 3.12 ([Cou90, CE12]). *For any integer $t \geq 0$ and any MSO_2 formula ϕ of length ℓ, an algorithm can be constructed which takes a graph G with treewidth at most t and decides in $O(f(t, \ell) \cdot (n + m))$ time whether $G \models \phi$ where the function f from this time bound is a computable function of t and ℓ.*

By Theorem 3.4 – as well as Proposition 8.5 of Wood and Telle [WT07] – we know that outer *k*-planar graphs have treewidth $O(k)$. Therefore, expressing outer *k*-planarity by an MSO_2 formula whose size is a function of *k* would mean that outer *k*-planarity could be tested in linear time. However, this task might be out of the scope of MSO_2. The challenge in expressing outer *k*-planarity in MSO_2 is that MSO_2 does not allow quantification over sets of pairs of vertices v_1, v_2 when v_1 and v_2 are not connected by an edge. Namely, it is unclear how to express a set of pairs that forms the circular order of vertices on the boundary of our convex drawing. However, if this circular order forms a *Hamiltonian cycle* in our graph, i.e., the given graph is closed, then we can indeed express this in MSO_2. With the edge set of a Hamiltonian cycle of our graph in hand, we can then ask that this cycle was chosen in such a way that the other edges satisfy either *k*-planarity or *k*-quasi-planarity.

Theorem 3.13. *Closed outer k-planarity can be expressed in MSO_2. Thus, closed and also full outer k-planarity can be tested in linear time.*

Theorem 3.14. *Closed outer k-quasi-planarity can be expressed in MSO_2.*

For a closed outer *k*-planar or closed outer *k*-quasi-planar graph *G*, we want to express that two edges *e* and e_i cross. To this end, we assume that there is a Hamiltonian cycle E^* of *G* that defines the outer face. We partition the vertices of *G* into three subsets *C*, *A*, and *B*, as follows: let set *C* contain the endpoints of *e*, whereas *A* and *B* are the two subgraphs on the remaining vertices connected using only edges of E^*. This partition divides the vertices of *G* into two special sets, one left and the other one right of *e*. For such a partition, e_i must cross *e* whenever e_i has one endpoint in *A* and one in *B*.

$$\text{CROSSING}(E^*, e, e_i) \equiv (\forall A, B, C)\big[\big(\text{VERTEX-PARTITION}(A, B, C)$$
$$\wedge (I(e, x) \leftrightarrow x \in C) \wedge \text{CONNECTED}(A, E^*) \wedge \text{CONNECTED}(B, E^*)\big)$$
$$\to (\exists a \in A)(\exists b \in B)[I(e_i, a) \wedge I(e_i, b)]\big]$$

Now the crossing patterns for closed outer *k*-planarity and closed outer *k*-quasi-planarity can be described using the formulas presented above as follows:

$$\textsc{Closed Outer } k\textsc{-Planar}_G \equiv (\exists E^*) \Big[\textsc{Hamiltonian}(E^*) \wedge$$

$$(\forall e)\Big[(\forall e_1, \ldots, e_{k+1})\Big[\Big(\bigwedge_{i=1}^{k+1} e_i \neq e \wedge \bigwedge_{i \neq j} e_i \neq e_j\Big) \to \bigvee_{i=1}^{k+1} \neg \textsc{Crossing}(E^*, e, e_i)\Big]\Big]\Big]$$

Here we insist that G is Hamiltonian and that, for every edge e and any set of $k+1$ distinct other edges, at least one among them does not cross e. The following formula directly implies Theorem 3.14.

$$\textsc{Closed Outer } k\textsc{-Quasi-Planar}_G \equiv (\exists E^*) \Big[\textsc{Hamiltonian}(E^*) \wedge$$

$$(\forall e_1, \ldots, e_k)\Big[\Big(\bigwedge_{i \neq j} e_i \neq e_j\Big) \to \bigvee_{i \neq j} \neg \textsc{Crossing}(E^*, e_i, e_j)\Big]\Big]$$

Again, we insist that G is Hamiltonian and further that, for any set of k distinct edges, there is at least one pair among them that does not cross.

This gives us linear-time recognition of closed outer k-planar graphs.

3.5 Conclusion

In this chapter, we explored two extended outerplanar settings – namely outer k-planarity and outer k-quasi-planarity.

For the outer k-planar graphs, we have shown that they are $(\lfloor\sqrt{4k+1}\rfloor+1)$-degenerate and thus also have chromatic number $(\lfloor\sqrt{4k+1}\rfloor + 2)$. We further showed that they have separators of bounded size – $2k + 3$, improving the old bound obtained from the bounded treewidth. This allowed us to give an algorithm for testing outer k-planarity in quasi-polynomial time.

For the outer k-quasi-planar graphs, we looked into comparability to planar graphs, showing that outer 3-quasi-planar graphs and planar graphs are incomparable under containment by providing graphs in either class but not in the other. We have also reconsidered the edge-maximal outer k-quasi-planar graphs, showing that those graphs are also edge-maximum. We obtained this result by giving a recursive argument, enumerating crossing patterns involving carefully picked edges.

In the last section, we provided an overview to Monadic Second-Order Logic. We used MSO_2 to express closed and full outer k-planar and outer k-quasi-planar graphs – insisting on the boundaries of the biconnected components to be Hamiltonian cycles for that component. Together with the bounded treewidth and Courcelle's Theorem, we showed that full and closed outer k-planarity can be tested in polynomial time.

Possible future research directions involve looking for polynomial-time algorithms to recognize outer k-planar graphs for $k \geq 2$ (since testing full outer k-planarity can be done in linear time) and outer k-quasi-planar graphs for $k \geq 3$ (as outer 2-quasi-planar graphs are exactly quasiplanar graphs).

As some of our claims were verified using a computer SAT solvers, we provide the formulations of our models in the next section. While we only used them to overcome exhaustive checking by hand, they are generally extendable and might be of interest for future research.

3.6 Additional Resources

3.6.1 Outer quasi-planarity checker

In this section, we describe a Boolean formula for testing whether a given graph is outer quasi-planar. We present the formula in first-order logic. After transformation to Boolean logic, we solve the formula using the MINISAT [SE05] solver.

A quasi outer-planar embedding corresponds to a circular order of the vertices. Cutting a circular order at some vertex v turns the circular into a linear order starting and ending at v. However, the edge crossing pattern remains the same. Therefore, we look for a linear order.

We need the following two sets of variables. For any pair of vertices $u \neq v \in V$, we introduce a Boolean variable $x_{u,v}$ that expresses that vertex u is before v in the linear order. In addition, for any pair of edges $e \neq e' \in E$ we introduce a Boolean variable $y_{e,e'}$ that expresses that edge e crosses edge e'. Now we list the sets of clauses present in our SAT formula.

$$x_{u,v} \wedge x_{v,w} \Rightarrow x_{u,w} \qquad \text{for each } u \neq v \neq w \in V; \qquad (3.5)$$

$$x_{u,v} \Leftrightarrow \neg x_{v,u} \qquad \text{for each } u \neq v \in V; \qquad (3.6)$$

$$x_{u,u'} \wedge x_{u',v} \wedge x_{v,v'} \Rightarrow y_{e,e'} \qquad \text{for each } e = (u,v) \neq e' = (u',v') \in E; \qquad (3.7)$$

$$\neg \left(y_{e_1,e_2} \wedge y_{e_1,e_3} \wedge y_{e_2,e_3} \right) \qquad \text{for each } e_1, e_2, e_3 \in E \text{ with different endpoints.} \qquad (3.8)$$

The first two sets of clauses describe the necessary properties of a linear order. The clause (3.5) realizes transitivity, and clause (3.6) anti-symmetry. Clause (3.7) ensures that variable $y_{e,e'}$ is set to true whenever the linear ordering on the endpoints of e and e' implies a crossing. Finally, clause (3.8) ensures that no three edges pairwise cross.

Chapter 4

One-Bend Drawings of Outerplanar Graphs with Fixed Shape

One of the fundamental problems in graph drawing is to draw a planar graph crossing-free under certain geometric or topological constraints. Many classical algorithms draw planar graphs under the constraint that all edges have to be straight-line segments, such as those by Schnyder [Sch90], de Fraysseix, Pach, and Pollack [dFPP90], or Tutte [Tut63]. But for practical applications we do not always have the freedom of drawing the whole graph from scratch, as some important parts of the graph may already be drawn. For example, in visualizations of large networks, certain patterns may be required to be drawn in a standard way, or a social network may be updated as new people enter a social circle or as new links emerge between already existing persons. In that case, we might want to extend a given drawing to a drawing of the whole graph.

For planar graphs, this problem is known as the PARTIAL DRAWING EXTENSIBILITY problem. Formally, given a planar graph $G = (V, E)$, a subgraph $H = (V', E')$ with $V' \subseteq V$ and $E' \subsetneq E$ called the *host* graph, and a planar drawing Γ_H of host H, the problem asks for a planar drawing Γ_G of G such that the drawing of H in Γ_G coincides with Γ_H. This problem was first proposed by Brandenburg et al. [BEG$^+$04] in 2003. Since then it has received a lot of attention in the subsequent years.

In this chapter, we consider a special drawing extension setting: We are given a biconnected outerplanar graph $G = (V_I \cup V_O, E_I \cup E_O)$. The edge set is divided into two subsets E_O and E_I: The edges in E_O define a Hamiltonian cycle in G that prescribes the outer face; the other set of edges E_I are the *inner* edges. Having the edge set partitioned into two subsets, we can partition the vertices in a similar way: Subset V_I contains all vertices incident to edges of E_I, whereas V_O contains all other vertices. As the *host graph*, we are given the subgraph $H = (V_I \cup V_O, E_O)$. The drawing Γ_H of H forms a simple polygon P in which all edges are to be drawn as straight-line segments. One can imagine the vertices of V_I being mapped to the boundary of P. We want to draw the missing edges of E_I inside the region defined by P such that Γ_G is crossing free while allowing one bend per edge of E_I.

A preliminary version of the contents of this chapter has appeared in the proceedings of EuroCG 2020 [AKL$^+$20]. This is joint work with Patrizio Angelini, Philipp Kindermann, Lena Schlipf, and Antonios Symvonis.

4.1 Related Work and Contribution

For the case of extending a given straight-line drawing using straight-line segments as edges, Patrignani [Pat06] showed the problem to be $\mathcal{NP}$-hard, but he could not prove membership in $\mathcal{NP}$, as a solution may require coordinates not representable with a polynomial number of bits. Recently Lubiw, Miltzow, and Mondal [LMM18] proved that a generalization of the problem where overlaps (but not proper crossings) between edges of $E \setminus E'$ and E' are allowed is hard for the existential theory of the reals ($\exists\mathbb{R}$-complete).

These results motivate allowing bends in the drawing. Angelini et al. [ADF$^+$15] presented an algorithm to test in linear time whether there exists any topological planar drawing of G with pre-drawn subgraph, and Jelínek, Kratochvíl, and Rutter [JKR13] gave a characterization via forbidden substructures. Chan et al. [CFG$^+$15] showed that a linear number of bends $(72|V'|)$ per edge suffices. This number is also asymptotically worst-case optimal as shown by Pach and Wenger [PW01] for the special case of the host graph not containing edges ($E' = \varnothing$).

Special attention has been given to the case that the host graph H is exactly the outer face of G. Already Tutte's seminal paper [Tut63] showed how to obtain a straight-line convex drawing of a triconnected planar graph with its outer face drawn as a prescribed convex polygon. This result has been extended by Hong and Nagamochi [HN08] to the case that the outer face is drawn as a star-shaped polygon without chords (that is, interior edges between vertices on the outer face). Mchedlidze, Nöllenburg, and Rutter [MNR13] give a linear-time algorithm to test for the existence of a straight-line drawing of G in the case that H is an arbitrary cycle of G and Γ_H is a convex polygon. Mchedlidze and Urhausen [MU18] study the number of bends required based on the shape of the drawing of H and show that one bend suffices if H is drawn as a star-shaped polygon.

Contribution. For any constant number k of bends, there exists some instance such that G has a k-bend drawing but no $(k-1)$-bend drawing; see, e.g., Figure 4.1 (b) for $k = 2$. Hence, it is of interest to test for a given k whether a k-bend drawing of G exists. This task is trivial for $k = 0$.

In this chapter, we introduce the ONEBENDINPOLYGON algorithm (see Algorithm 4.1), which solves this problem for $k = 1$ in time $O(|V_I| \cdot p)$, where p is the number of corners of P. In Section 4.2 we establish notation and the lemmata necessary to describe the algorithm. We then state ONEBENDINPOLYGON in Section 4.3; finally in Section 4.4 we prove correctness and runtime.

4.2 Notation and Preliminaries

In the following, we consider the setting described above, but with the notation simplified as follows: We summarize all elements of host graph H under the polygon P such that the Hamiltonian cycle coincides with the boundary ∂P. Hence, the problem at hand is equivalent to finding an outerplanar one-bend drawing of the subgraph $G_I = (V_I, E_I)$

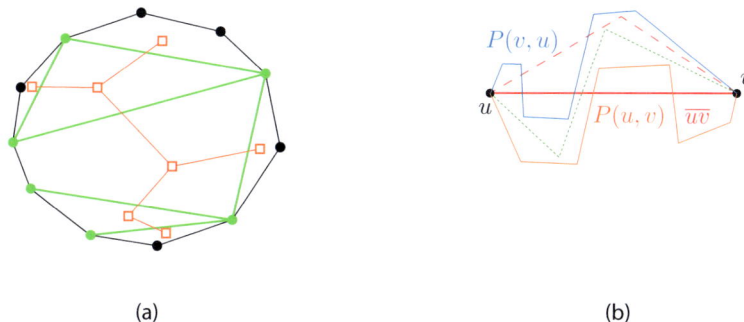

Figure 4.1: (a) A biconnected outerplanar graph, subgraph G_I in green and the dual tree in orange. (b) For an edge $e = (u, v)$, the straight line $\overline{uv}$ intersects both $P(u, v)$ and $P(v, u)$, the dashed red 1-bend drawing of e only avoids crossing $P(u, v)$, possible 2-bend drawing in green.

inside P where the vertices V_I are already mapped to ∂P. We say that G *can be drawn in P* if there is a crossing-free drawing of G with its vertices on ∂P as defined by the mapping, its outer face drawn as ∂P, and its interior edges drawn with at most one bend per edge.

For a pair of vertices u and v mapped to ∂P, we denote by $\overline{uv}$ the straight-line segment between them. The given mapping also orders the vertices of V_I along the boundary. Starting at u and following ∂P in counterclockwise order until reaching v, we obtain the open interval $P(u, v)$ – the piece of ∂P between u and v. As a complement, we also have the open interval $P(v, u)$ – the piece between v and u. By concatenation of the two vertices and the two pieces we get $\partial P = u \circ P(u, v) \circ v \circ P(v, u)$.

Considering the full graph G, the faces of G induce a unique dual tree T (e.g. see Proskurowski and Syslo [PS86]) where each edge of T corresponds to an *interior* edge of E_I; see Figure 4.1 (a). Any interior edge $e = (u, v)$ divides the polygon into two parts and for an edge corresponding to a leaf of the dual tree, one of these parts does not contain other vertices of V_I. The faces corresponding to these empty parts are exactly the leaves of T. In the following, we consider T to be rooted at some leaf node f_n. For each face f_i, we denote by $p(f_i)$ the parent of f_i in T, and by e_i the edge between f_i and its parent. We say that f_i and f_j are *siblings* if $p(f_i) = p(f_j)$.

ONEBENDINPOLYGON will traverse the dual tree twice – first bottom-up and then top-down. We can consider the faces of G in sequence of the bottom-up traversal $f_1, \ldots, f_n$. In each step, we incrementally process an interior edge of G_I, and prune T and refine P accordingly. We will now define the pruning and refinement operations to then give a description of the algorithm.

The sequence in which the edges are processed also implies a sequence of subtrees of T. For step i (with $1 \le i \le n$), let T_i be the subtree of T induced by the nodes $f_i, \ldots, f_n$.

Hence for the first step we have $T_1 = T$ and for the final step we have $T_n = (\{f_n\}, \varnothing)$. By pruning the tree, eventually every node of T will become a leaf node of some subtree.

Similar to the sequence of pruned subtrees, we also have a corresponding sequence of refined polygons $P_1, \dots, P_n$. In the first step we have $P_1 = P$ and after the last step of the bottom-up traversal P_n is the bounding cycle of f_n. In step i, we process the edge $e_i \in E_I$ and by choice of sequence, one of the parts induced by e_i in P_i will be a leaf corresponding to f_i in T_i. Putting a one-bend drawing of e_i into P_i using bend point b, we classify the type of corner that b will induce in P_{i+1}. We say that the interior edge e_i is either

- a *reflex* edge if bend point b has to be placed in a way that enforces a reflex corner to occur at b in P_{i+1} – see Figure 4.2 (a) – or

- a *convex* edge if a placement of b resulting in a convex corner at b in P_{i+1} or as a straight line is possible.

Let G_i be the subgraph of G_I induced by the vertices incident to the faces $f_i, \dots f_n$, hence $G = G_1$. In step i, our algorithm picks a leaf f_i of T_i and processes the interior edge e_i between f_i and its parent such that the following invariant holds:

"Graph G_{i+1} can be drawn in P_{i+1} if and only if G_i can be drawn in P_i."

To maintain this invariant during the bottom-up traversal of T, we want to refine P in the least restrictive way – cutting away as little of P as possible. We will formally prove that the invariant holds in Lemma 4.6.

4.3   Procedure

The objective for this section is to establish the pieces needed to ONEBENDINPOLYGON (Algorithm 4.1). Before we can do that, we describe how it will proceed in more detail, establishing the necessary lemmata on the way.

Among all leaves of T_i, the algorithm chooses the next node f_i to process as follows: If T_i has a leaf corresponding to a reflex edge, we process the corresponding interior edge next. Otherwise, all leaves in T_i correspond to convex edges, and we choose one of the nodes of the dual tree among them that has the largest distance to the root f_n in T. We do this to make sure that a convex edge is only chosen if all siblings corresponding to reflex edges have already been processed.

Let $e_i = (u, v)$ be the interior edge corresponding to the leaf f_i. Let V_u and V_v be the regions inside of P_i visible from u and v, respectively, and let $V_{e_i} = V_u \cap V_v$ be their intersection – the region visible by both end points. Clearly, any valid bend point for e_i needs to be inside V_{e_i}. For any point $b \in V_{e_i}$, let $Q^b_{e_i}$ be the *obstructed region*, the subpolygon of P_i bounded by $P_i(u, v) \circ \overline{vb} \circ \overline{bu}$ – the part of P_i that is "cut off" by drawing e_i with its bend at b. We call a bend point $b \in V_{e_i}$ *minimal* for e_i if there is no other point $b' \in V_{e_i}$ with $Q^{b'}_{e_i} \subsetneq Q^b_{e_i}$.

For reflex edges, we have the following lemma regarding minimal bend points.

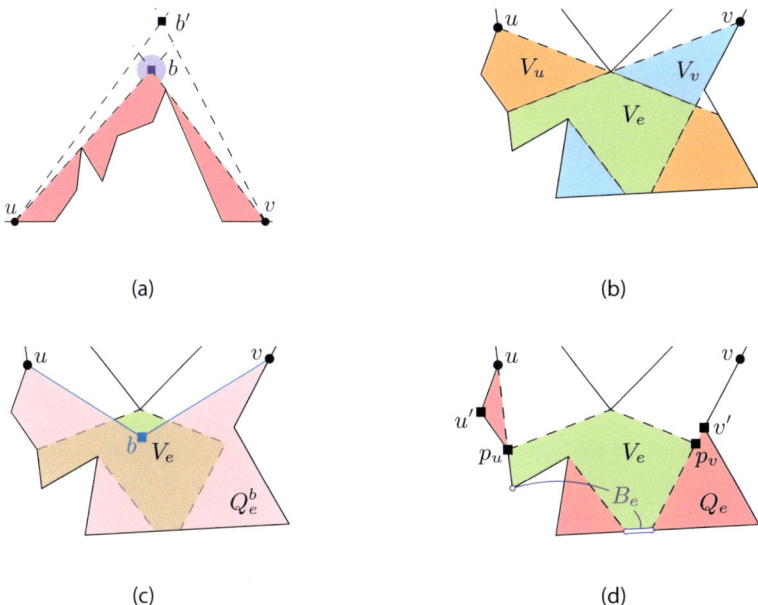

(a)

(b)

(c)

(d)

Figure 4.2: Illustrations of minimal bend points and obstructed regions for an edge e connecting u and v: (a) Edge e is reflex: We show two possible drawings using either b or b': Both cut away at least the red region and b is the minimal bend point to do so. In both cases, the modified polygon has a reflex angle at the bend. (b) Edge e is convex: Visibility region V_u of vertex u in orange, visibility region V_v of vertex v in blue, and intersection $V_e = V_u \cap V_v$ in green; (c) the region Q_e^b cut off by drawing e with its bend at point b in red; (d) Construction of p_u, p_v and of the obstructed region Q_e, set of bend point options B_e in purple.

Lemma 4.1. *Let $e_i = (u, v)$ be a reflex edge to be drawn in polygon P_i. If there is a valid drawing of e_i in P_i, then there is also a unique minimal bend point b for e_i.*

Proof. We construct b by considering the shape of $P(u, v)$ and the visibility region V_{e_i}. As e_i is a reflex edge, some parts of $P(u, v)$ must extend over $\overline{uv}$ – either by intersecting $\overline{uv}$ or by $\overline{uv}$ being completely outside of P_i.

By assumption there is some valid drawing of e_i; let b' be the bend point of that drawing; see Figure 4.2 (a). Consider the line $\overline{ub'}$ and rotate it clockwise around u until it intersects $P(u, v)$. Rotating $\overline{ub'}$ towards v also moves the intersection point $b^\dagger$ of both lines closer to v. That way, we obtain the obstructed region $Q_{e_i}^{b^\dagger} = P(u, v) \circ \overline{ub^\dagger} \circ \overline{b^\dagger v}$ and $Q_{e_i}^{b^\dagger} \subset Q_{e_i}^{b'}$. Doing the same with line $\overline{b^\dagger v}$, rotating counterclockwise towards u, we get the intersection point b of both rotated lines.

Moving intersection point b by a small constant distance ε to avoid intersecting $P(u,v)$ with either line segment, it becomes a valid bend point for e_i and in total we have $Q_{e_i}^b \subset Q_{e_i}^{b^\dagger} \subset Q_{e_i}^{b'}$. As ε decreases towards 0, the point b becomes a minimal bend point for edge e_i by construction. $\qquad\square$

Considering convex edges, we can no longer rely on having a single minimal bend point. Hence, we need to refine our notation.

Given edge e_i, let B_{e_i} be the set of all valid minimal bend points. We define the region $Q_{e_i} = \bigcap_{b \in B_{e_i}} Q_{e_i}^b$ to be the region of P_i *obstructed* by all valid drawings e_i – wherever we place the bend point of e_i, all the points of Q_{e_i} will be cut off. Conversely, for every point $p \in P_i \setminus Q_{e_i}$, there is a placement of the bend point of e_i such that p is not cut off. An example for the obstructed region and the set of possible minimal bend points for a convex edge can be seen in Figure 4.2 (d).

Lemma 4.2. *Let e_i be a convex edge to be drawn in polygon P_i. If there is a valid drawing of e_i in P_i, then $B_e \subset \partial V_e$ and we can safely refine P_i by removing Q_e.*

Proof. For any reflex edge, the unique minimal bend point is the point on the boundary of that edge's visibility region defining the smallest obstructed region (Lemma 4.1). When considering a convex edge e_i, the boundary ∂V_{e_i} and $P(u,v)$ can coincide in some points or even segments. Our goal is to preserve as much of V_{e_i} as possible for future usage while still refining the polygon. Therefore, we first define the set B_{e_i} of all bend points with obstructed regions that are incomparable with respect to containment. Then we compute the region Q_{e_i} obstructed by all these bend points.

Let (u, u') and (v', v) be the segments of $P_i(u,v)$ incident to u and v, respectively. Consider the ray starting at u and going towards u', coinciding with $\overline{uu'}$. Rotate this ray in counterclockwise direction until it hits V_{e_i} for the first time; call this point p_u. Do the same with the ray from v to v', rotating it clockwise; let the point where it hits V_{e_i} be p_v. To identify all points belonging to B_{e_i}, consider the outline piece of the visibility region between p_u and p_v – namely $V_{e_i}(p_u, p_v)$. The piece $V_{e_i}(p_u, p_v)$ is composed of two things: Rays with origin u or v that are tangent to ∂P_i, and parts of ∂P_i itself. The set B_{e_i} is composed of all points on $\partial P_i \cap V_{e_i}(p_u, p_v)$ that are not also on some tangent ray, and of those points on tangent rays that have the largest distance to the origin of that ray.

We now describe how to construct Q_{e_i} using p_u and p_v as intersecting the individual obstructed regions for all possible bend points is infeasible. Using the points and segments constructed above we get that Q_{e_i} is the region inside P_i bounded by the following segments:

$$Q_{e_i} = \overline{vp_v} \circ V_{e_i}(p_u, p_v) \circ \overline{p_u u} \circ P_i(u,v).$$

Notice that this region is not necessarily simple, as tangent rays and parts of $V_{e_i}(p_u, p_v)$ can coincide with parts of ∂P_i.

If there is an edge e_j later in the sequence ($i < j$) of G_I that needs to have its bend point inside Q_{e_i} then any two drawings of e_i and e_j have to intersect and G_I cannot be drawn in P. Thus, cutting away Q_{e_i} is a safe refinement of P_i. $\qquad\square$

The set of bend points B_e constructed for edge e in Figure 4.2 (d) contains a single point (to the left), as it is the end point of tangent rays with origins u and v; there is also a segment of $P_i \subset B_e$ – including the segments' end points, as they are endpoints of tangent rays.

The complete pseudocode for ONEBENDINPOLYGON is listed in Algorithm 4.1:

4.4 Correctness

We now proceed to show correctness of ONEBENDINPOLYGON (Algorithm 4.1). As described above, it consists of two while-loops: the first one represents the bottom-up traversal whereas the second one represents the top-down traversal. Processing the edges and safely refining the polygon accordingly is treated in Lemma 4.1 and Lemma 4.2, respectively: In step i of the first loop, our algorithm computes the visibility region V_{e_i}. If we have a step with an empty visibility region, it is impossible to draw G_i in P_i, so by the invariant it is also impossible to draw G_I in P and the algorithm stops. Otherwise, the algorithm computes Q_{e_i} and creates $P_{i+1} = P_i \setminus Q_{e_i}$. If an edge e_j ($j > i$) had to place its bend point inside Q_{e_i}, then e_i and e_j would cross independent of the choice of the bend point of e_i; in this case, our algorithm concludes that it is impossible to draw G in P when processing e_j.

In the following, we focus on the top-down traversal: We show that G_{i+1} can be drawn in P_{i+1} if and only if G_i can be drawn in P_i.

We first analyze the possible sequences in which sibling-sets consisting of convex edges can be processed. Without being able to fix the "best" drawing of a convex edge, we need to argue that there is no bad sequence for picking the nodes of convex edges. To do so, for each convex edge e_i, we define the region $R_{e_i} = \left(\bigcup_{b \in B_{e_i}} Q^b_{e_i} \right) \setminus Q_{e_i}$ to be the region of P_i *restricted* by e_i – see the red region in Figure 4.3 (a). For each point $r \in R_{e_i}$, there are two minimal bend points b and b' for e_i such that bending e_i at b cuts off r, whereas bending e_i at b' does not.

Lemma 4.3. *Let $S(f)$ be the set of all nodes in T with parent f that correspond to convex edges. For any pair of edges $e_1, e_2 \in S(f)$ and polygon P_i to draw inside, the restricted regions R_{e_1} and R_{e_2} are interior-disjoint.*

Proof. Let $e_1 = (u_1, v_1)$ and $e_2 = (u_2, v_2)$ be two convex edges with parent node f and let f correspond to edge $e = (u', v')$. Consider the three pieces $P(u', v'), P(u_1, v_1)$ and $P(u_2, v_2)$ defined by the parent and its two children. Since e_1 and e_2 are siblings below f, they are "next to" each other along $P(u', v')$ – that is, w.l.o.g. we get $P(u_1, v_1) \subset P(u', v')$ and $P(u_2, v_2) \subset P(u', v')$ but $P(u_1, v_1) \cap P(u_2, v_2) = \varnothing$. s Since both edges are convex, their visibility regions – and thus also their restricted regions – live between $P(u', v')$

Algorithm 4.1: OneBendInPolygon(Outerplanar Graph G_I, Polygon P)

Tree $T \leftarrow$ dual of G_I with outer face P
Stack $\mathcal{C} \leftarrow \varnothing$ /* to store convex edges */
Set $\mathcal{L} \leftarrow$ Set of leaves in T
Set $\mathcal{B} \leftarrow \varnothing$ /* to store edge-bend point pairs */
node $f_n \leftarrow$ node from $\mathcal{L}$ /* as root for T */
node $f_i \leftarrow$ Nil, region $Q_e \leftarrow$ Nil   /* iteration variables */
while $(\mathcal{L} \setminus f_n) \neq \varnothing$ **do**   /* bottom-up traversal */

> **if** $\mathcal{L}$ *contains leaf nodes corresponding to reflex edges* **then**
>
> > pick $f_i \in \mathcal{L}$ for some reflex edge e
> > **if** $V_e = \varnothing$ **then return** Impossible
> > compute optimal bend point $b \in V_e$ and region Q_e /* Lemma 4.1 */
> > $\mathcal{B}.\text{Add}(\{e, b\})$
>
> **else**   /* $\mathcal{L}$ only contains convex edges */
>
> > pick $f_i \in \mathcal{L}$ for a convex edge $e = (u, v)$ /* safe by Lemma 4.3 */
> > **if** $V_e = \varnothing$ **then return** Impossible
> > compute minimal bend points B_e and regions Q_e /* Lemma 4.2 */
> > $\mathcal{C}.\text{Push}(\{e, P(u, v)\})$ /* for top-down traversal */
>
> $P_i \leftarrow P_i \setminus Q_e$ /* refine P */
> $T \leftarrow T \setminus f_i$ /* prune T */
> Remove f_i from $\mathcal{L}$, check if parent $p(f_i)$ is now a leaf

while $\mathcal{C} \neq \varnothing$ **do**   /* top-down traversal */

> $\{e = (u, v), P_{\text{old}}(u, v)\} \leftarrow \mathcal{C}.\text{Pop}()$
> $P \leftarrow u \circ P(v, u) \circ v \circ P_{\text{old}}(u, v)$ /* re-expand P, Lemma 4.4 */
> recompute minimal bend points B_e
> **if** $B_e \cap P = \varnothing$ **then return** Impossible   /* later refinement */
> point $b \leftarrow$ bend point from $B_e \cap P$
> $\mathcal{B}.\text{Add}(\{e, b\})$
> $P \leftarrow P \setminus Q_e^b$ /* refine P */

return $\mathcal{B}$

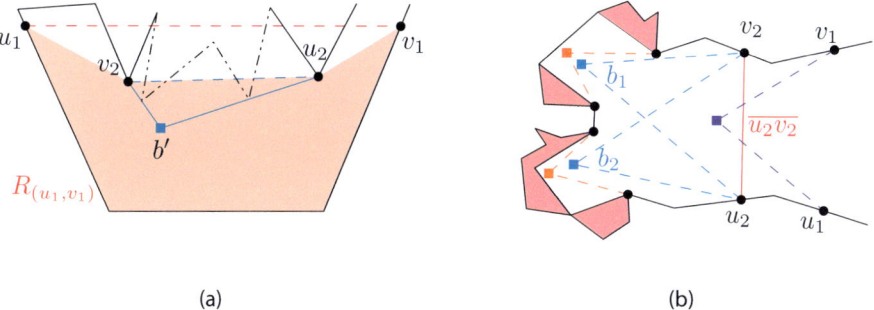

(a)   (b)

Figure 4.3: (a) The dashed part of piece $P(u_2, v_2)$ forces the bend point b' to be placed inside $R_{(u_1,v_1)}$. To do so, b' creates a reflex angle, making (u_2, v_2) a reflex edge that would have been processed before (u_1, v_1). (b) Edge $e_1 = (u_1, v_1)$ is the parent of edge $e_2 = (u_2, v_2)$. The two possible bend points b_1, b_2 of edge e_2 are each placed in the restricted region of one of the children of e_2. Fixing the one-bend drawing of e_1 to intersect $\overline{u_2 v_2}$ makes e_2 become a reflex edge, eliminating any choice.

and $\overline{u_1 v_1}$ or $\overline{u_2 v_2}$, respectively; see Figure 4.3 (a). W.l.o.g. assume that $\overline{u_2 v_2}$ is between $\overline{u_1 v_1}$ and $P(u', v')$ as seen along the boundary. Hence, the structures of $P(u', v')$ that u_2 and v_2 are mapped to influence the shape of the visibility region V_{e_1} – in the worst case $\overline{u_2 v_2}$ is part of boundary ∂V_{e_1}. The only reason for e_2 to be drawn in a way that its bend point would be inside R_{e_1} is that some part of $P(u_2, v_2)$ intersects $\overline{u_2 v_2}$ – see the dashed piece $P(u_2, v_2)$ in Figure 4.3 (a). This would imply that e_2 is a reflex edge, a contradiction. $\qquad\square$

While bend points for reflex edges can be fixed immediately, convex bends have to remain undecided until the root of the dual tree is reached. Given a convex edge e_i encountered in step i of the bottom-up traversal and some later edge e_j (with $i < j$), two events can have an impact on how e_i will be drawn:

- If e_j is a reflex edge – hence encountered later in the bottom-up traversal – it might require its unique optimal bend point to be placed inside R_{e_i}, or

- if e_j is a convex edge, it will be fixed before e_i in the top-down traversal. The bend point of e_j might need to be placed inside R_{e_i} – see edge (u_2, v_2) in Figure 4.3 (b).

All restricted regions can possibly be subject to further refinement until encountered again during the top-down traversal. To reconstruct the situation of step i while incorporating all refinements, for any convex edge $e_i = (u, v)$, we store the piece $P_i(u, v)$ on stack $\mathcal{C}$. When f_n is eventually encountered, all safe refinements of nodes below it have been removed from P, creating the final polygon P_n; it incorporates all refinements necessary to draw any reflex edges and also the obstructed region of the last convex edge

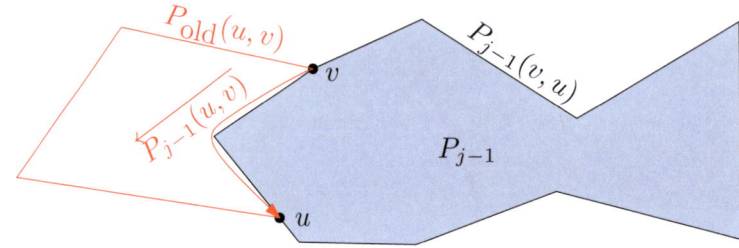

Figure 4.4: Constructing polygon P_j for edge (u, v) from P_{j-1} by adding the stored piece $P_{old}(u, v)$.

encountered. This obstructed region might in turn contain the visibility regions of other yet to be fixed convex edges. Assuming that we encountered and stored c convex edges, we create c additional *expanded* polygons $P_{n+1}, \ldots, P_{n+c}$; for each stored convex edge (u, v), we create the expanded polygon P_j from P_{j-1} by replacing $P_{j-1}(u, v)$ with the stored piece $P_{old}(u, v)$. During the bottom-up traversal, we kept the part of P_i bound to containing root f_n; in the top-down traversal, this is also reversed – when given a choice, we keep the part of the polygon containing the next convex edge e_j – in Algorithm 4.1 this edge is stored as $C.\text{Top}$. We get the following lemma:

Lemma 4.4. *Let $(e_j = (u, v), P_{old}(u, v))$ be the next convex edge-piece pair to process in the top-down traversal and P_{j-1} be the current refined polygon. We can obtain the simple polygon P_j to draw e_j into by replacing $P_{j-1}(u, v)$ with $P_{old}(u, v)$.*

Proof. We first prove that the current polygon's boundary ∂P_{j-1} contains both vertices u and v. Then we show that P_{j-1} can safely be expanded to P_j.

If there is any sub polygon containing only one of the two vertices, there must have been some edge separating them – with u on one side and v on the other. Then this edge and (u, v) must cross in any outer drawing, hence contradicting outerplanarity of the input graph. Hence, if they are both cut off of P_{j-1}, both must have been cut off by edge e' at the same time. This edge then separates the polygon into two pieces, both having a boundary partially composed of the drawing for e' and with one of them containing u and v; choose the boundary of this subpolygon as ∂P_{j-1}.

The vertices u and v subdivide boundary of P_{j-1} into two pieces – see Equation (4.1).

$$P_{j-1} = u \circ P_{j-1}(u, v) \circ v \circ P_{j-1}(v, u) \tag{4.1}$$

$$P_j = u \circ P_{old}(u, v) \circ v \circ P_{j-1}(v, u) \tag{4.2}$$

$$= P_{j-1} \cup u \circ P_{old}(u, v) \circ v \circ \underbrace{P_{j-1}(u, v)}_{\text{reversed}} \tag{4.3}$$

To obtain P_j from P_{j-1}, we replace piece $P_{j-1}(u, v)$ with the stored piece $P_{old}(u, v)$ as in Equation (4.2). By processing the nodes of dual tree T in the order described above we know that all refinements that happened between the step when edge e was stored and the current step j are either part of piece $P_{j-1}(v, u)$ or cut off by that piece. Any refinement made inside piece $P_{old}(u, v)$ – not cutting it off completely – would contradict that the node corresponding to edge (u, v) being a leaf at the time. To see that P_j is free of self-intersections, notice that we expanded P_{j-1} by effectively joining it with the region described in Equation (4.3); any segment of $P_{j-1}(v, u)$ intersecting $P_{old}(u, v)$ must also have intersected $P_{j-1}(u, v)$. This joining operation is illustrated in Figure 4.4.

With P_j being a simple polygon and edge $e = (u, v)$ being a leaf in the corresponding dual tree, we can now recompute the visibility region V_e. $\qquad\square$

During step j of the top-down traversal ONEBENDINPOLYGON will check and (possibly) draw edge e_j without looking back or ahead. In Lemma 4.4 we established that not looking back is safe. For not looking ahead we have the following lemma.

Lemma 4.5. *Let $e_j = (u, v)$ be the convex edge to be drawn in P_j, let V_{e_j} be the visibility region of e_j, and let $S(e_j)$ be the set of all (not yet fixed) convex children of e_j in T. If there is a valid drawing for all edges in $S(e_j)$ inside their corresponding polygon before refining it by drawing e_j, then there still are valid drawings for all edges after drawing e_j.*

Proof. The edges in $S(e_j)$ are siblings in T; hence, by Lemma 4.3, their respective restricted regions are area-disjoint. Pick a valid bend point b for e_j and assume that it is in the restricted region of edge $e' \in S(e_j)$. As e_j has only one bend point to place and as the restricted regions are disjoint, e' is the only edge affected by the choice of b – see edge (u_2, v_2) with two different possible bend points b_1 and b_2 in Figure 4.3 (b). Notice that b cannot be in the obstructed region $Q_{e'}$ as this region was subject to refinement, is not part of the current polygon P_j and hence cannot be part of V_{e_j}. Next consider the possible drawing of e' using bend point b'; assume that placing b cuts off b' from P_j and thus also from the polygon corresponding to e'. By the definition of the restricted region $R_{e'}$ for e' we know that $b' \in R_{e'}$ and therefore we can find an alternative bend point $b^* \in R_{e'}$ to bend edge e' at later. $\qquad\square$

Combining the results above yields the following lemma.

Lemma 4.6. *If the visibility region for the current edge is non-empty, then G_{i+1} can be drawn in P_{i+1} if and only if G_i can be drawn in P_i.*

Proof. During the bottom-up traversal, we refine the current polygon, only making it smaller in every step. Hence, whenever we have $V_{e_i} = \varnothing$ for any edge e_i inside P_i, we stop as refining further cannot increase the size of any visibility regions. Otherwise, we either compute the unique minimal bend point b (Lemma 4.1), or the obstructed region Q_{e_i} as described in Lemma 4.2, safely refining P_i to P_{i+1} accordingly. Since minimal bend points cannot lie in restricted regions of siblings (Lemma 4.3), only the bend point of the edge corresponding to $p(e_i)$ can possibly be placed in the restricted region R_{e_i} of e_i; therefore, any edge drawn in the bottom-up traversal is correct and safe.

During the top-down traversal, we can encounter overlapping restricted regions, but only when the nodes are in a parent-child relationship (Lemma 4.3). Then by Lemma 4.5, when there is a bend point placement for the current edge e_j that is valid for the current refined polygon, this choice cannot impact the decisions later in the traversal; that is because any other bend point that lies inside $R(e_j)$ must lie on the opposite side of the drawing of $p(e_j)$ (Lemma 4.4), so it cannot influence the choice of the bend point for e_j.

$\square$

We are now ready to state the main result of this chapter.

Theorem 4.7. *Given an outerplanar graph $G_I = (V_I, E_I)$, a polygon P with p corners, and a mapping of the vertices of G_I to the boundary ∂P, we can decide in $O(|V_I| \cdot p)$ time whether G_I can be drawn in P with at most one bend per edge.*

Proof. We use ONEBENDINPOLYGON as described in Algorithm 4.1. The correctness follows immediately from Lemma 4.6.

The most expensive part of the algorithm in terms of runtime is to compute the visibility region V_e for all the edges $e = (u, v)$. Since V_e is a simple polygon with at most $2p$ edges, it can be computed in $O(p)$ time, as demonstrated by Gilbers [Gil14, page 15]. Doing two traversals, the visibility region of each edge needs to be computed at most twice; as outerplanar graphs can have at most an amount of edges linear in the number of vertices, all these regions can be computed in $O(|V_I| \cdot p)$ total time. The remaining parts of the algorithm (computing the dual graph of G, choosing the order of the faces f_i in which we traverse the graph, computing Q_e, "cutting off" parts of P, and propagating the graph at the end to fix the presentation) can clearly be done within this time.

Thus, the total running time is $O(|V_I| \cdot p)$. $\square$

4.5 Conclusion

In this chapter, we have developed an algorithm that (when possible) is able to draw an outerplanar graph into a predefined simple polygon when it is allowed to add one bend per edge. When no such drawing exists, our algorithm reports the edges that enforce the crossing. This can then be used to identify the malformed segments of the boundary – forcing the bend points to be outside of the polygon, enforcing the crossing – using the dual tree of the drawing. We phrased the problem as a partial representation extension problem, where the outer face is a Hamiltonian cycle that is pre-drawn. The task then was to add the missing chords, bending each missing edge at most once and only inside the drawing.

This technique was designed with the application in mind that a fixed drawing of some planar subgraph is already given that then needs to be completed, respecting the shapes of the prescribed faces. Considering that the mapping of the vertices to the boundary is part of its input, our algorithm can be used to accomplish this goal – or report that no such drawing extension exists.

Open problems on this chapter could be to consider drawing related graph classes: What about planar graphs with only a subset of the vertices mapped to the boundary, or outer 1-planar graphs? Another direction would be to consider other pre-drawn structures, such as a path or (more generally) a tree? If the pre-drawn path is colinear, the problem becomes testing for a 2-page book embedding with a fixed layout.

Acknowledgments. This work was initiated at the Workshop on Graph and Network Visualization 2019. We thank all the participants for helpful discussions and Anna Lubiw for bringing the problem to our attention.

Part II

Vertices on the Integer Coordinates

Chapter 5

Moving Graph Drawings to the Grid Optimally

From a graph drawing perspective, restricting the vertex coordinates to be of integer precision can be desirable for various reasons, such as aesthetics, computational complexity, or technical limitations. While the $\mathcal{NP}$-hardness proof (Section 5.2) and the integer linear programming formulation (Section 5.3) were already submitted and presented as part of a Master's Thesis [Löf16], we include these results in this work for the sake of completeness: In the following two chapters, we consider the problem TOPOLOGICALLY-SAFE GRID REPRESENTATION. This chapter covers the theoretical results on that problem and presents an exact algorithm that finds optimal solutions to an $\mathcal{NP}$-hard problem. This motivates the results we will present in Chapter 6. With a geographic application in mind – representing road networks and administrative borders at finite precision –, we use this chapter as a foundation to design and evaluate a randomized heuristic approach to solving TOPOLOGICALLY-SAFE GRID REPRESENTATION.

Concepts. In the computational geometry community, a process called *snap rounding* has been proposed for line arrangements and has since become well-established. Let the euclidean plane be tiled into unit squares called pixels with center on integer coordinates. Let S be a finite collection of line segments $s \in S$ in the plane and let $\mathcal{A}(S)$ be the *arrangement* of vertices, edges and faces in the plane induced by the segments and intersections of S. Guibas and Marimont [GM98] define the snap rounding paradigm:

Definition 5.1. *Snap rounding* is the process of converting the arbitrary precision arrangement $\mathcal{A}(S)$ into a fixed-precision representation $\mathcal{A}^*(S^*)$ with these properties:

(1) Fixed-precision: All vertices of $\mathcal{A}^*$ are at integer precision coordinates.

(2) Geometric similarity: For each segment $s \in S$, the transformed segment s^* lies within the Minkowski sum of s and a pixel at the origin.

(3) Topological similarity: $\mathcal{A}$ and $\mathcal{A}^*$ are "topologically equivalent up to the collapsing of features" – that is, there is a continuous deformation of the segments in S to their snap-rounded counterparts such that no segment ever passes completely over a vertex in the arrangement.

A preliminary version of the contents of this chapter has appeared in the proceedings of Graph Drawing 2016 [LvDW16]. This is joint work with Thomas C. van Dijk and Alexander Wolff [Löf16].

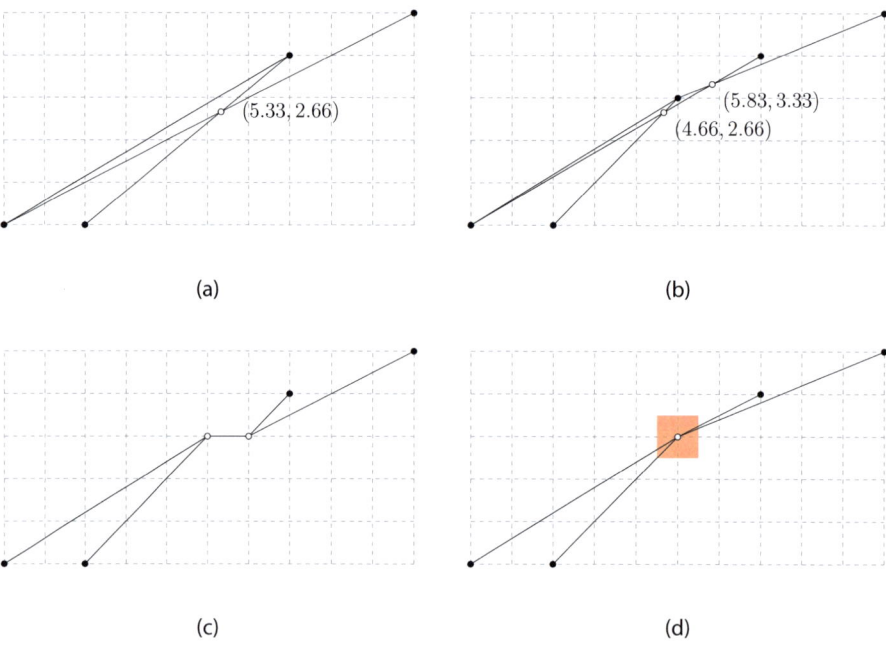

Figure 5.1: Rounding the intersection point of two line segments: (a) An example instance on four vertices and three edges, intersection point depicted as a white vertex, (b) rounding the intersection to the nearest integer grid point creates two extraneous intersections. The results obtained by using the algorithms of (c) Greene and Yao and (d) Hobby (tolerance square in red).

Designed to overcome problems induced by working with infinite-precision real arithmetic machines (RAMs) – a paradigm used by Preparata and Shamos [PS85], de Berg, Halperin, and Overmars [dBHO07], and others – , this technique can also be used for limited display resolutions, such as bitmap graphics. There are several algorithms for computing such a representation that are fast and work well in practice; we present a brief survey on challenges and solutions in Section 5.1.

The concepts common to all approaches stem from the line intersection problem as stated by Greene and Yao [GY86]: Given an arrangement of line segments, each pixel that contains vertices or intersections is called *hot*. Then every segment becomes a polygonal chain whose edges (*fragments*) connect center points of hot pixels, namely those that the original segment (*ursegment*) intersects. Guibas and Marimont [GM98] showed that during snap rounding, vertices of the arrangement never cross a polygonal chain, so after snapping no two fragments cross. Moreover, the circular order of the fragments around an output vertex is the same as the order in which the corresponding ursegments intersect the boundary of its pixel. The resulting arrangement approximates the original one in the sense that any fragment lies within the Minkowski sum of the corresponding ursegments and a unit square centered at the origin. But by definition, the output of those algorithms is not topologically safe: vertices, edges or even faces can visually disappear from the output drawing while rounding.

5.1   Related Work and Contribution

5.1.1   Rounding to the Grid

Greene and Yao [GY86] considered the precision required to store the points created when intersecting line segments as it has numerous applications in computational geometry. Reducing the precision used to store these intersection points can lead to artificial *extraneous intersections*, that are not part of the original arrangement – see Figure 5.1. Initially, these intersections have been handled by repeatedly running an intersection detection algorithm – for example the Bentley–Ottmann sweep [BO79] –, until no new intersections are reported. The algorithm given by Greene and Yao does not prevent extraneous intersections from occurring, but finds and rounds all intersections in one single iteration. An example output of their algorithm can be found in Figure 5.1 (c).

The approach by Hobby [Hob99] considers *tolerance squares* – unit-length pixels centered on integer grids in which segment endpoints and intersections occur. Edges passing through tolerance squares get subdivided, then every vertex in the square gets snapped to the center. While this may introduce new incidences, it avoids extraneous intersections – see Figure 5.1 (d).

Guibas and Marimont [GM98] give a boiled down definition of snap rounding and state a dynamic algorithm based on vertical cell decompositions; Having n unrounded segments, a set $H = \{h_1, \ldots, h_{|H|}\}$ of tolerance squares, complexity of the cell decomposition C and complexity of the arrangement A, their algorithm has an output-sensitive asymptotic runtime of $O(n \log n + A + \sum_{h \in H} |h|^2 + C)$.

Goodrich et al. [GGHT97] give two simplified algorithms: The first algorithm is deterministic and based on the Bentley-Ottmann sweep, the other algorithm is randomized using trapezoidal decomposition. Both have a matching runtime of (expected) $O(n \log n + \sum_{h \in H} |h| \log n)$, independent of arrangement complexity A.

The precision required to store vertex coordinates and to measure vertex-to-vertex distances can be bounded by grid resolution, but measuring distances between nonincident vertex-edge pairs may still require arbitrary precision. Rounding the endpoints of segments induces *drift* on the segment itself. Drift can cause a rounded segment to pass through a tolerance square it did not pass in its unrounded state; this is demonstrated by edge e_1 in Figure 5.2 (a) and (b). To overcome this, Halperin and Packer [HP02] augment the classic snap rounding procedure by iterating the process. This *iterated snap rounding* procedure gives an output that is equivalent to that of repeatedly applying the tolerance square-based rounding process until for all pairs of vertices and nonincident edges, the distance is at least half the width of a pixel. Applying this to the above example

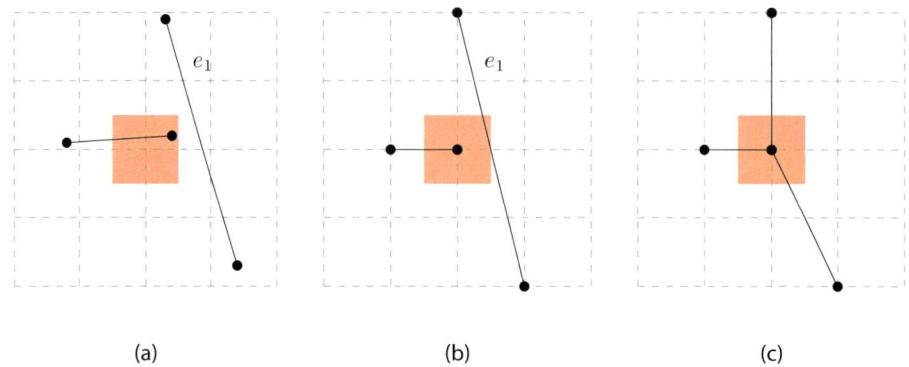

Figure 5.2: Iterated snap rounding: (a) Input segments at arbitrary precision, edge e_1 does not intersect the red tolerance square. (b) Edge e_1 intersects the tolerance square after snap rounding, requiring another iteration. (c) Resulting arrangement after iterated snap rounding, edge e_1 subdivided.

(Figure 5.2 (c)) edge e_1 gets subdivided during the second iteration and both parts subsequently become incident to the snapped vertex. Subdividing segments and rounding the pieces can imply additional drift induced by consecutively intersecting other tolerance squares, heavily deforming the output arrangement. Effort to bound the drift was made by Packer [Pac06], adding a user-specified parameter. An implementation of iterated snap rounding for $2D$ arrangements can also be found in the CGAL computational geometry framework [Pac19].

De Berg, Halperin, and Overmars [dBHO07] extend the original list of desired properties adding *non-redundancy*. A degree 2 vertex of the output is redundant if it does not stem from a segment endpoint; consider the white vertex in Fig 5.3 (b). They give an algorithm that outputs a snap-rounded arrangement in which all redundant vertices are removed by using a second vertical sweep. This algorithm has a total runtime of $O((n + I)\log n)$ with I being the number of intersections in the arrangement.

Hershberger [Her13] introduced *stable snap rounding*. Algorithms based on tolerance squares can be made sable in the following sense: the rounded arrangement does not change when re-applying the procedure. This is obtained by providing individual rules for two different types of tolerance squares, requiring $O(|H|\log n)$ additional runtime.

Most recently, rounding of arrangements in 3D has recently been studied by Devillers, Lazard, and Lenhart [DLL18]. Rounding specific classes of graphs, such as Voronoi diagrams (studied by Devillers and Gandoin [DG02]) have also been considered from a computational geometry perspective and with similar objective in mind.

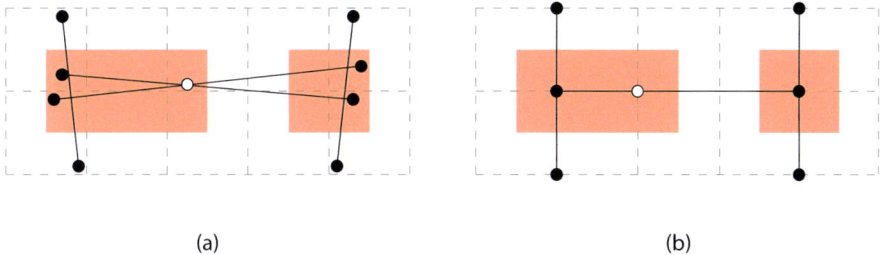

(a)   (b)

Figure 5.3: Redundant degree 2 vertices in rounded arrangements: (a) input segments (redundant intersection marked as white vertex), (b) output with (white) degree 2 vertex that is not a segment endpoint.

5.1.2   Drawing on the Grid

From a graph-drawing perspective, restricting vertex coordinates to integer precision is a common practice. Fáry [Fár48] (among others) shows that every planar graph has a planar straight line embedding with vertices as points on the plane; this is also known as a *Fáry embedding*. Tutte [Tut63] introduces the barycenter method for drawing planar graphs. It yields drawings that need precision linear in the size of the graph.

Dolev, Leighton, and Trickey [DLT83] introduce the family of planar *nested triangles graphs* – see Figure 5.4. Nested triangles with n vertices can be used to prove that $(2n/3 - 1) \times (2n/3 - 1)$ is a lower bound on the required area when restricting straight line drawings on the integer grid.

Motivated by these results, Schnyder [Sch90] and, independently, de Fraysseix, Pach, and Pollack [dFPP90] have shown that any planar graph with n vertices admits a straight-line drawing on a grid of size $O(n) \times O(n)$ and that this is asymptotically optimal in the worst case. Chrobak and Nakano [CN98] have investigated drawing planar graphs on grids of smaller width, at the expense of a larger height.

Krug and Wagner [KW07] give a reduction from 3-PARTITION, showing that area minimization of straight line grid drawings is $\mathcal{NP}$-hard. They also give an iterative algorithm that computes a more compact drawing of a given plane graph.

Nöllenburg and Wolff [NW11] give a mixed integer program for octilinear metro-map drawings with station labels. They establish sets of hard and soft constraints to create a visually pleasant map drawing for answering navigational questions while not preserving real-world distances or travel times. While solving a very special problem, their model be adapted for other geometric tasks – we will do this in Section 5.3.

In relation to that, Biedl et al. [BBN+13] propose a generic ILP model for various grid-based layout problems, such as determining pathwidth, finding optimum s-t-orientations or bar k-visibility representations.

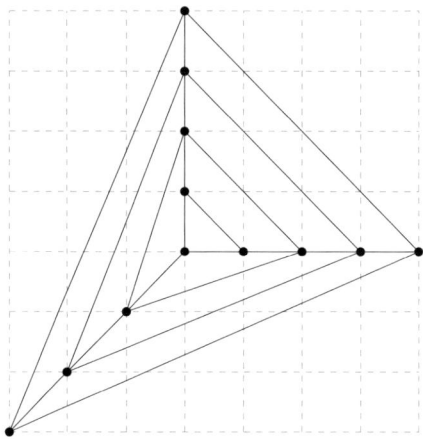

Figure 5.4: Nested triangles graph on $n = 12$ vertices.

Contribution.   We investigate the problem of moving the given drawing of a planar graph to a given grid, prioritizing the ability to recognize the original graph over geometric similarity: While we do not tolerate new point-point or point-line incidences in the rounded drawing, we accept the possibility that a vertex does not go to the nearest grid point. Presented this set of requirements, our objective is to minimize the change induced on the vertex positions. This change can by measured, for example, by the sum of the distances or the maximum distance in the Euclidean (L_2-) or Manhattan (L_1-) metric. We define our variant of this problem – Topologically-Safe Grid Representation – in Section 5.2 and show that it is $\mathcal{NP}$-hard; To do so, we give a reduction that asks for compressing each coordinate by just a single bit. The proof is somewhat similar in concept to the proof of the $\mathcal{NP}$-hardness of Metro-Map Layout by Nöllenburg [Nöl05], but requires additional constructions since the rounding problem does not easily allow the construction of "rigid" gadgets. In Section 5.3, we propose an integer linear program (ILP) to solve instances of Topologically-Safe Grid Representation to optimality; Our ILP formulation generalizes the ILP for Metro-Map Layout by Nöllenburg and Wolff [NW11] in the following way: Where the ILP by Nöllenburg and Wolff assumes a constant number of possible edge directions (namely 8) to obtain an octilinear drawing, our desired output asks for a number of edge directions that is quadratic in the size of the grid; on a grid of size $k \times k$, there are $\Theta(k^2)$ edge directions. The numbers of variables and constraints of our ILP are polynomial in grid size and graph size, but are quite large in practice. Thus, for an n-vertex planar graph, we must generate $O(k^2n^2)$ constraints, among others, to preserve planarity and the cyclic order of edges around the vertices. To ameliorate this, we apply delayed constraint generation, a technique used to add certain constraints only when needed. However, the runtime of our ILP is prohibitive for graphs with more than about 15 vertices – we give a set of example-instances in $2D$ in Section 5.4 to show the power and limitations of our exact approach. Our techniques can also be adapted to draw (small) graphs with minimal area. This is interesting even for small graphs since minimum-area drawings can be useful for validating (counter)examples in graph drawing theory.

5.2 NP-Hardness

In the following we will consider a rounding task that relaxes on geometric similarity of the output while enforcing topological equivalence; see properties (2) and (3) of Definition 5.1. We first give a formal definition of this new problem that we call TOPOLOGI-CALLY-SAFE GRID REPRESENTATION and then proceed to show $\mathcal{NP}$-hardness.

Definition 5.2 (TOPOLOGICALLY-SAFE GRID REPRESENTATION). As *input* we take a plane graph $G = (V, E)$ with vertex positions of arbitrary precision and a bounding rectangle $B = [0, X_{max}] \times [0, Y_{max}]$.

The TOPOLOGICALLY-SAFE GRID REPRESENTATION problem is trying to find an drawing Γ of G the with the following properties:

(1) All vertices of G are moved to integer coordinates that are contained inside B, and

(2) Γ is topologically equivalent to the given plane straight-line drawing of G.

(3) The total *displacement* of Γ is minimal over all such drawings: the displacement of a single vertex is the Manhattan-distance between its original position and the rounded coordinate, and the displacement of a drawing is the the sum of over all vertex displacements.

We prove $\mathcal{NP}$-hardness of TOPOLOGICALLY-SAFE GRID REPRESENTATION by considering the decision variant. Instead of searching for the smallest total displacement, we ask for a drawing with some constant displacement c. Having an algorithm to answer this question, we can use it to search for the smallest such constant c_{min}. Since the searching can be done using a polynomial amount of queries, proving hardness for the decision variant also implies hardness of the original problem.

We reduce from PLANAR MONOTONE 3SAT. A formula F for this variant of 3SAT (recall from Section 2.2.2) has the following additional properties:

- The graph $H(F)$ induced by the formula – using variables and clauses as vertices and having an edge for every occurrence – is *planar*.

- Clauses are *monotone*: The variables of a clause are all negated or all unnegated.

The PLANAR MONOTONE 3SAT problem is known to be $\mathcal{NP}$-hard, as shown by de Berg and Khosravi [dBK12]. We can assume that the graph $H(F)$ can be laid out as shown in Figure 5.5 (a): all vertices corresponding to variables lie on the x-axis with the vertices of the all-negated clauses above them and the vertices of all-unnegated clauses below them.

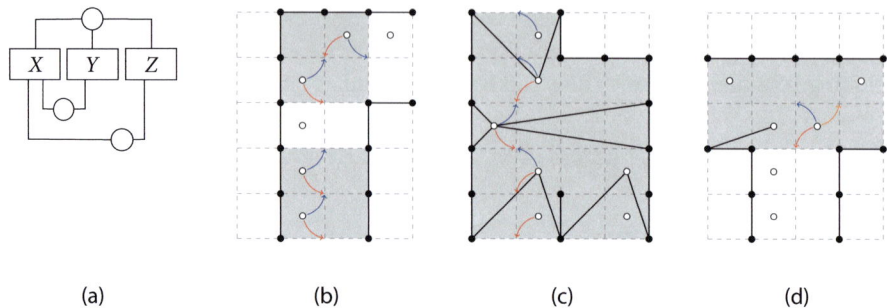

(a)   (b)   (c)   (d)

Figure 5.5: (a) Graph $H(F)$ for formula $F = (\overline{X} \vee \overline{Y} \vee \overline{Z}) \wedge (X \vee Y) \wedge (X \vee Z)$; (b–e) All gadgets used in the reduction. Inner area of each gadget highlighted gray, possible roundings indicated by arrows. (b) Vertical line gadget (bottom) connected to corner gadget (top), (c) variable gadget (with two negated and one unnegated occurrences), and (d) all-negated clause gadget with three negated variables.

Theorem 5.1. TOPOLOGICALLY-SAFE GRID REPRESENTATION *is* $\mathcal{NP}$-*hard*.

Proof. For a formula F of PLANAR MONOTONE 3SAT, we construct a cost bound c_{min} and a plane graph G with vertices at half-integer coordinates[1]. The sum of all vertex movements induced by rounding G to integer coordinates will be exactly c_{min} if and only if F is satisfiable. To achieve this, we introduce gadgets to resemble the elements of the formulas incidence graph $H(F)$ – variables, clauses, edges and bends – and construct G and c_{min} in polynomial time.

For exposition, we draw the vertices of G using two different styles. Black vertices start on integer grid points and do not need to be rounded. Moving a black vertex to another integer grid point is allowed, but we will show that this is not optimal if F is satisfiable. White vertices start at grid cell centers – using half-integer coordinates – and thus will always move at least 1 unit – rounding each coordinate by at least $|\pm 0.5|$ independently. Let $W \subseteq V(G)$ be the set of white vertices. Now we give the construction of the various gadgets.

To get started, we introduce the line and bend gadgets. They are used to consistently transport the variable assignments to the clause gadgets. Every segment of the line gadget consists of four black vertices and two edges forming a *tunnel*, and a single white vertex inside; see Figure 5.5 (b), lower half. Each white vertex can be rounded most cheaply – at cost 1 – to exactly two possible integer grid points inside the tunnel, depicted by the red and blue arrows. Consider the white vertices of two neighboring grid cells inside a tunnel. They share exactly one grid point that is not occupied by black vertices of the tunnel's wall. Rounding one of the white vertices to that grip point prevents the other vertex from also going there; with only one safe option of cost 1 left, the other white

---
[1] Storing half-integers requires an additional bit compared to regular integers.

vertex has to mimic the movement of the first vertex. So, if the white vertex at one end of a tunnel is rounded inward (blue arrow) the white vertex at the other end of that line must be rounded outward – we say it is *pushed*. The same machinery works for the bend gadgets, as can be seen in Figure 5.5 (b), upper half.

Next, consider the variable gadget depicted in Figure 5.5 (c). It can be extended horizontally to have tunnels for vertical line gadgets for every negated and unnegated occurrence at the top and bottom respectively. Inside the left-most of the center cells of this gadget, there is a white *assignment* vertex. The assignment vertex is connected to the gadget's walls by four edges forming two triangles and can be rounded up or down representing the two possible states of the variable. Rounding the assignment vertex to either corner of its cell forces a pair of those triangle edges to coincide with a line of the grid, blocking grid points on the top or bottom tunnels, respectively. The tunnels connected on that side of the gadget are then all forced to push inwards and into the connected clause gadgets. Following the layout convention depicted in Figure 5.5 (a), we say that a variable is set to true, if the assignment vertex is rounded upwards (following the blue arrow) and false, if rounded downwards (following the red arrow).

Finally, the clause gadget is shown in Figure 5.5 (d). In the following, we only discuss the all-negated degree-3 version of this gadget. The degree-2 version can be constructed by replacing the vertical tunnel by a black vertex at the tunnels entrance and extending the walls on both sides, connecting them to the new black vertex to form a tunnel. The all-unnegated versions of this gadget can be obtained by mirroring the all-negated versions at a horizontal line. At the center of the gadget, there is a white *satisfaction* vertex that can go to any of three possible integer grid points (or two respectively) at equal cost. These grid points belong to the three line gadgets and are only available if the corresponding line does not transmit a push. Then the satisfaction vertex can be rounded at cost 1 if and only if the clause is satisfied.

All white vertices must be rounded at cost at least 1. Thus, the rounding cost of G is bounded from below by $c_{min} = |W|$. If F is satisfiable, there is a rounding of all vertices of W to achieves this: Round the assignment vertices according to a satisfying assignment of the variables. Then the line and corner gadgets between a clause and the (at least) one variable satisfying it does not transmit a push. This allows the corresponding satisfaction vertex to be rounded towards the entrance of that line. All rounded white vertices contribute cost of exactly 1. In the other direction, a satisfying assignment can be read off from the assignment vertices if rounding occurred at cost c_{min}.

For all three candidate grid points of a satisfaction vertex to be unavailable, all line gadgets connected to the clause must be forced to transmit pushes; hence, none of the variables occurring in this clause are assigned to satisfy this clause. In our construction, this shows as three variable-tunnel-clause triples. In each such triple, all integer points are occupied by white vertices, leaving no grid point for the satisfaction vertex of the clause. A topologically valid rounding of such a triple then must involve moving a black vertex, making the original position of the black vertex available by broadening the tunnel. Since all white vertices still need to be moved by at least 1, adding the cost of moving the black vertex makes G exceed the cost bound c_{min}. That is, if c_{min} is exceeded, then F

is unsatisfiable: Any rounding corresponding to a satisfying truth assignment is cheaper. This concludes our reduction and the claim follows. □

Given that the objective function we intend to optimize is polynomially bounded by the size of the bounding rectangle and the number of vertices in the instance, the following extension of Theorem 5.1 implies that there is no fully polynomial-time approximation scheme unless $\mathcal{P} = \mathcal{NP}$ [GJ79].

Corollary 5.2. TOPOLOGICALLY-SAFE GRID REPRESENTATION *is $\mathcal{NP}$-hard in the strong sense.*

Proof. The only numerical variables in an instance of TOPOLOGICALLY-SAFE GRID REPRESENTATION are vertex coordinates. In the proof of Theorem 5.1 the constructed instances have a bounding rectangle of polynomial size and thus coordinate values are limited polynomially as well. Thus the runtime of a hypothetical algorithm remains exponential in input size when unary representations are used. □

We decided to prove $\mathcal{NP}$-hardness of TOPOLOGICALLY-SAFE GRID REPRESENTATION using the Manhattan distance as a cost measure, because of the integer linear program we present in Section 5.3. Linearizing Manhattan distance with standard transformations [MS97], it can easily be used as an objective function. The hardness result itself is not limited to these considerations.

Corollary 5.3. TOPOLOGICALLY-SAFE GRID REPRESENTATION *is also $\mathcal{NP}$-hard when using Euclidean distance to evaluate rounding costs. In this case it is also $\mathcal{NP}$-hard to minimize the maximum movement d_{max} instead of the sum over all vertices.*

Proof. The choice of distance measure changes the lower bound c_{min}. Using Euclidean distance, rounding a white vertex costs at least $\sqrt{0.5^2 + 0.5^2}$ and moving black vertices still costs (at least) 1. Thus, we obtain a minimum rounding cost of $c_{min} = \sqrt{0.5^2 + 0.5^2} \cdot |W|$. As moving black vertices is still not accounted for, the above proof still holds.

Exploiting the fact that $\sqrt{0.5^2 + 0.5^2} \approx 0.71 < 1$, we can identify satisfiable instances by considering the maximum movement d_{max} over all vertices of G. Given an instance of TOPOLOGICALLY-SAFE GRID REPRESENTATION, a maximum movement of $d_{max} < 1$ implies that only white vertices have been rounded. Thus, the corresponding formula is satisfiable. For the case of an unsatisfiable formula, at least one black vertex needs to be moved, making $d_{max} \geq 1$. □

The ability to make the distinction between different maximum movements (moving a black vertex at cost 1 versus only moving white vertices at cost 0.71) based on the satisfiability of F also gives the following.

Corollary 5.4. *Euclidean* TOPOLOGICALLY-SAFE GRID REPRESENTATION *with the objective to minimize maximum movement d_{max} is $\mathcal{APX}$-hard.*

5.3 Exact Solution Using Integer Linear Programming

In this section we provide an exact algorithm for TOPOLOGICALLY-SAFE GRID REPRE-SENTATION by giving an integer linear program – or ILP for short. In the following, we describe the integer variables necessary for instances of TOPOLOGICALLY-SAFE GRID REPRESENTATION to then model the task of finding an optimal solution to TOPOLOGI-CALLY-SAFE GRID REPRESENTATION by giving a linear cost function and several sets of linear constraints to ensure topological equivalence. Our model borrows ideas from a linear program for creating octilinear drawings of metro maps by Nöllenburg and Wolff [NW11], changing and extending them to fit for our requirements.

In the following presentation, we will use upper-case letters to denominate constants, whereas lower-case letters will be variables for the ILP. We further divide the variables, greek characters will be used for binary variables while variables represented by roman characters can take any natural number as value.

5.3.1 Basic Model

Each instance of TOPOLOGICALLY-SAFE GRID REPRESENTATION is a graph $G = (V, E)$ together with real-valued vertex coordinates $p: V \to \mathbb{R}^2$ and a prescribed rotation system. In our model, we consider the input coordinates of each vertex $v \in V$ to be a tuple of real-valued constants (X_v, Y_v). We define the integer-valued coordinates $r: V \to \mathbb{N}^2$ with $r(v) = (x_v, y_v)$, thus obtaining two integer variables $0 \le x_v \le X_{\max}$ and $0 \le y_v \le Y_{\max}$ to represent the rounded output coordinates in our model. The ILP will find an optimal solution by adjusting the values of these variables. This data model naturally leads to the objective function of Equation (5.1).

$$\text{Minimize} \sum_{v \in V} |x_v - X_v| + |y_v - Y_v| \tag{5.1}$$

By using the absolute value function, Equation (5.1) is not linear, but can be made so with standard transformations [MS97]. Note that without any further constraints, this would just move vertices to (one of) the nearest integer grid points. Together with a polynomial-time checker for topological equivalence, this already implements the most basic heuristic. We will refer to this heuristic as *Instant Rounding* and build on this idea in Section 5.3.3 – and also later in Section 6.2 of the next chapter (see page 88).

5.3.2 Constraints for Topological Equivalence

Unique Vertex Coordinates. Given the representations of coordinates in our model, ensuring that vertices do not coincide can easily be done by adding the constraints of Equation (5.2). They too are not linear as stated, but can also be readily linearized.

$$(x_v \ne x_w) \lor (y_v \ne y_w) \qquad \forall v, w \in V, v \ne w \tag{5.2}$$

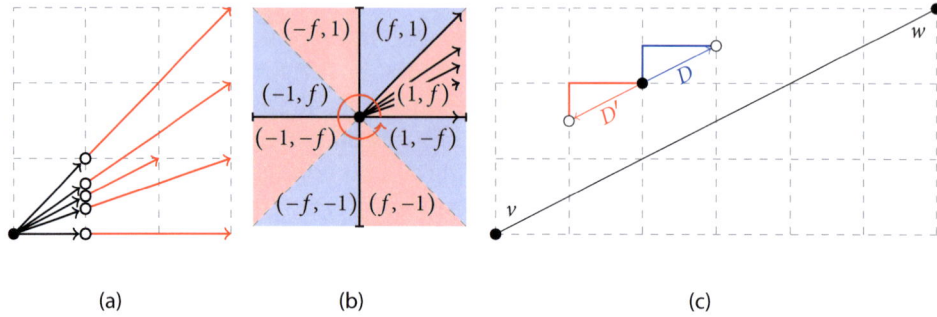

(a)   (b)   (c)

Figure 5.6: Farey sequence and direction assignment: (a) The slopes of the black vectors match the elements of depth 3 in the Farey sequence $\{0, 1/3, 1/2, 2/3, 1\}$, the red extensions point to all grid points of distance at most three within that plane octant. (b) Using the elements of the sequence as coordinates for vector endpoints (either as x or y coordinate and with different signs), we get sets of vectors for all octants; combining all sets, we get set $\mathcal{D}$. Programmatically creating $\mathcal{D}$, we can order all directions radially around the origin. (c) Edges have two directions assigned to them – one for each endpoint. Direction D (blue) matches the slope of (v, w) as seen from v while D' (red) matches the slope seen from w. D and D' are opposites, which is also represented in the ordering of $\mathcal{D}$.

We will now introduce constraints to ensure that input and output are topologically equivalent; that is, no two edges intersect and the edges at every vertex have the same cyclic order as in the input. To do this, we first introduce the following tool:

Possible Directions. The most important departure from the metro-map drawing ILP is about the number of different edge slopes. The goal of Nöllenburg and Wolff was to draw metro maps in the rather classic octilinear style. This restriction is unreasonable for planar graphs in general, more than eight different directions are required. A priori the only assumption on the placement of rounded vertices is that they each lie on a grid point somewhere within the given bounding rectangle. Hence, edges in a valid drawing can possibly go from any of those grid points to any other. Let $\mathcal{D}$ be the set of unique directions $D = (D_X, D_Y)$ in $[-X_{\max}, X_{\max}] \times [-Y_{\max}, Y_{\max}]$. To explicitly enumerate all elements of this set, we use the Farey sequence [GKP94]. The Farey sequence F_n recursively enumerates all fully-reduced fractions with denominator less or equal to n. More precisely, F_n contains F_{n-1} and all mediants[2] created from subsequent elements of F_{n-1}; the first three elements of the Farey sequence are shown in Equation (5.3).

$$F_1 = \left\{\frac{0}{1}, \frac{1}{1}\right\}, F_2 = \left\{\frac{0}{1}, \frac{1}{2}, \frac{1}{1}\right\}, F_3 = \left\{\frac{0}{1}, \frac{1}{3}, \frac{1}{2}, \frac{2}{3}, \frac{1}{1}\right\} \qquad (5.3)$$

---

[2] Also known as the "freshman sum", the mediant of two fractions $\frac{a}{b}$ and $\frac{c}{d}$ is defined as the sum of the numerators divided by the sum of the denominators $\frac{a+c}{b+d}$.

Picking $n = \max\{X_{max}, Y_{max}\}$, we use the n-th element of the sequence F_n to generate the set $\mathcal{D}$ of all possible directions as follows: We use the elements in $f_1, \ldots, f_k \in F_n$ as slopes for a set of vectors; all vectors start at the origin and the j-th vector points towards coordinate $(1, f_j)$. Extending these vectors to travel a distance of n in x-direction, they together point to (or pass over) all grid points that are inside the bounding rectangle and the first octant. An illustration for $n = 3$ can be found in Figure 5.6 (a). Alternating the order and signs of the endpoint's coordinates allows us to create sets of vectors for all eight octants; see Figure 5.6 (b). From the literature [GKP94] we get that $|F_n| \approx \frac{3n^2}{\pi^2}$ and hence the number of all possible directions inside the bounding rectangle is $|\mathcal{D}| \in \Theta(n^2)$. In the following, we let the set $\mathcal{D}$ be ordered counterclockwise and enumerated using integers, starting at the positive x-axis, allowing comparison of directions by integer identifier.

No two Edges Cross. The following constraints ensure that nonincident edges do not cross without preventing incident edges from touching in their shared vertex. We will follow the general idea of Nöllenburg and Wolff [NW11]. While producing octilinear drawings of metro maps, they ensured planarity by forcing every pair of nonincident edges to be separated by at least some distance D_{min} in at least one of the eight octilinear directions. Enforcing this minimum distance of separation was partly an aesthetic guideline, but also guarantees planarity. We employ the same approach for the latter, but as we are not interested in the same aesthetics, some modifications are necessary. Most notably, we allow for more than eight possible edge slopes and thus need to check if a pair of nonincident edges is separated in any direction of $\mathcal{D}$. Enforcing a rather large minimum distance between edges was useful for Nöllenburg and Wolff when labeling the individual stations along the metro lines. We do not have additional labels that we need to reserve space for, therefore we pick the separation distance D_{min} such that all planar realizations on the grid are allowed; that is, D_{min} has to be small enough to separate any non-intersecting pair of edges in the output. Considering the possible slopes of an edge and the minimum distance of that edge to the endpoint of some other edge, we choose $D_{min} = 1/(\max\{X_{max}, Y_{max}\} + 1)$ – smaller than all non-zero elements in F_n.

For every pair of nonincident edges $e_1, e_2 \in E$ and all directions $D \in \mathcal{D}$, we add a binary variable $y_D(e_1, e_2) \in \{0,1\}$ to the model. A value of 1 indicates that e_1 and e_2 are apart by D_{min} in direction D. Every pair of nonincident edges must be separated in some direction (following the idea of Nöllenburg [Nöl05]), we need to make sure that at least one of the corresponding y is set to 1. Hence, we get the set of constraints shown in Equation (5.4).

$$\sum_{D \in \mathcal{D}} y_D(e_1, e_2) = 1 \qquad \forall e_1, e_2 \in E, e_1, e_2 \text{ nonincident} \qquad (5.4)$$

To enforce separation of the edges, we impose constraints on the coordinates of the edges' endpoints. We need all constraints to be contained in the model before solving, but as there are opposing directions, we also need to include sets of constraints that are apparently conflicting. Hence we only want the set of constraints to be enforced for which the corresponding y is set to 1. To model this switch-case scenario, we extend all necessary equations by a term composed of a multiplication of that γ by some large constant $L_y{}^3$, adding some slack to the constraints we want to have "disabled" during solving. Let $L_y = 2 \cdot \max\{X_{\max}, Y_{\max}\} + 1$. Then we require the following for any direction $D \in \mathcal{D}$, all pairs of nonincident edges e_1, e_2, and all pairs of endpoints $v \in e_1, w \in e_2$.

$$D_X \cdot (x_v - x_w) + D_Y \cdot (y_v - y_w) + (1 - \gamma_D(e_1, e_2))L_y \geq D_{\min}$$
$$\forall D \in \mathcal{D} \quad \forall e_1, e_2 \in E, e_1, e_2 \text{ nonincident} \quad \forall v \in e_1, w \in e_2 \tag{5.5}$$

Considering a single pair of nonincident edges, the constraint of Equation (5.4) yields a unique direction D with $\gamma_D = 1$. By choice of L_y, the constraints of Equation (5.5) that involve a direction D with $\gamma_D = 0$ are trivially fulfilled.

Determine Direction of Incident Edges. To prevent two incident edges $e_1, e_2 \in E$ from overlapping, we again generalize the metro-map drawing ILP. Without being restricted to the octilinear drawing style, we assign any direction from $\mathcal{D}$ to either edge. The direction assigned to an edge $e = (v, w)$ is relative to the endpoint of that edge – consider the opposing directions D and D' in Figure 5.6 (c). By ensuring that the directions assigned to any pair of incident edges differ, we prevent those edges from overlapping.

Following the recipe for nonincident edges, we introduce a binary decision variable $\alpha_D(v, w) \in \{0, 1\}$ for every vertex $v \in V$, every vertex $w \in N(v)$ from the neighborhood of v and every direction $D \in \mathcal{D}$. Setting $\alpha_D(v, w)$ to 1 implies that the direction of edge (v, w) is D when considered from endpoint v. To ensure that every edge gets some direction assigned to it, we use the constraints presented in Equation (5.6).

$$\sum_{D \in \mathcal{D}} \alpha_D(v, w) = 1 \qquad \forall v \in V \quad \forall w \in N(v) \tag{5.6}$$

For any vertex $v \in V$, any neighbor $w \in N(v)$, and any direction $D \in \mathcal{D}$, the following ensures that edge (v, w) indeed has direction D and that the position endpoint w matches that direction. Again presented with sets of constraints apparently in conflict, we add conditional slack to the inequalities as we did in Equation (5.5). We do so by using constant $L_\alpha = 2 \cdot \max\{X_{\max}, Y_{\max}\} + 1$.

$$x_w \cdot D_Y + y_v \cdot D_X - x_v \cdot D_Y + (1 - \alpha_D(v, w))L_\alpha \geq y_w \cdot D_X$$
$$x_w \cdot D_Y + y_v \cdot D_X - x_v \cdot D_Y - (1 - \alpha_D(v, w))L_\alpha \leq y_w \cdot D_X$$
$$(1 - \alpha_D(v, w))L_\alpha + (x_w - x_v) \cdot D_X + (y_w - y_v) \cdot D_Y \geq 0 \tag{5.7}$$
$$\forall v \in V \quad \forall w \in N(v) \quad \forall D \in \mathcal{D}$$

By setting some α to 1 – and thus satisfying the constraints of Equation (5.6) – we enable one subset of constraints from Equation (5.7), as L_α dominates all other terms. These constraints enforce the comparison between edge slope and direction, which gives us the direction of edge (v, w) with the correct sign.

With the correct value assigned to each α, we prevent overlapping incident edges and preserve cyclic orders of neighbors around each vertex in one set of constraints.

---

[3] In Operations Research, this is known as the *Big-M method*.

Preserve Cyclic Orders of Outgoing Edges.   We test cyclic orders using the mapping of the corresponding edges onto the directions, as those are already ordered linearly by the angle between the corresponding vector and the positive x-axis. Given the input embedding, we enforce that for vertex v and the i-th neighbor w_i of direction D_i, the direction D_{i+1} of the next neighbor w_{i+1} must be later in the order of $\mathcal{D}$, hence we say $D_i < D_{i+1}$ identifier-wise. We also need to linearize the cyclic order of neighbors around each vertex in order to be able to encode them into the model. This imposes the problem that we cannot know in advance, which neighbor will be assigned the direction with the lowest identifier. Given k neighbors and knowing the "correct" first neighbor in advance, we would get the ordering of the directions $D_1 < D_2 < \cdots < D_k$ to match $w_1, w_2, \ldots, w_k, w_1, \ldots$. Unrolling the cyclic order of neighbors at the wrong point, there could then be a neighbor in the linearized order, for which the increasing-identifier condition described above cannot hold, despite of the cyclic order being preserved in the output. To overcome this, we use a binary decision variable $\beta(v, w) \in \{0,1\}$ for every vertex-neighbor pair, indicating if w is the "last" neighbor of v according to the order of $\mathcal{D}$. Thus, when the output contains an ordering $D_1 < \cdots < D_\ell \not< D_{\ell+1} < D_{\ell+2} \ldots$, we use the corresponding β to add the slack necessary to disable exactly one constraint.

$$\sum_{w \in N(v)} \beta(v, w) = 1 \qquad \forall v \in V, \deg(v) > 1 \tag{5.8}$$

$$\alpha_{D_1}(v, w_i) \leq \beta(v, w_i) + \sum_{D_w \in \mathcal{D}: D_w > D_1} \alpha_{D_w}(v, w_{i+1})$$
$$\forall D_1 \in \mathcal{D} \quad \forall v \in V, N(v) = \{w_1, w_2, \ldots, w_k\}, k = \deg v > 1 \tag{5.9}$$

For notational convenience, we let $w_{k+1} = w_1$, as $N(v)$ is conceptually circular. For any α set to 0, the inequalities of (5.9) are trivially satisfied. Otherwise, there has to be a neighbor whose connecting edge has direction with higher identifier (and thus the corresponding α set to 1), unless it is the last neighbor in the embedding of v. To ensure that there is only one "last neighbor"-violation of the constraints from (5.9), we introduce the constraints of (5.8). Adding β to every constraint of (5.9) also allows for the whole neighborhood of v to be rotated around it.

Theorem 5.5. *The above ILP solves* TOPOLOGICALLY-SAFE GRID REPRESENTATION.

In addition to its original purpose, the integer linear program we described above can also be repurposed to produce grid drawings of planar graphs of small area. To do so, changes to the objective function are required; see Equation (5.10).

$$\text{Minimize } \max_{v \in V} y_v \tag{5.10}$$

Lemma 5.6. *By replacing the objective function (Equation (5.1)) from the above integer linear program with that shown in Equation (5.10), the above model can be used to search for minimal-area drawings.*

Proof. Replacing the objective function with that of Equation (5.10), the ILP computes a planar straight-line grid drawing with the given embedding and width at most $X_{\max}$

(prescribed by the bounding rectangle). For each now constant width, we get a drawing of minimal height by scaling down the graphs original coordinates to make it fit into one grid cell and then "rounding it" back to the grid. Changing the bounding rectangle width, we get an algorithm to search for a drawing of smallest bounding rectangle. □

5.3.3   Delayed Constraint Generation

The integer linear program described to solve TOPOLOGICALLY-SAFE GRID REPRESENTATION works in theory; however, it is not suited for practical applications due to obstructive computation time. This problem becomes even more apparent when trying to find drawings of small area. Hence, we discuss the application of *delayed constraint generation* (also known in the context of *constraint generation*, see Chinneck [Chi08]) to the original model as well as the area minimization variant. Delayed constraint generation is a technique used to speed up the process of solving linear programs. We discuss its usefulness and give an experimental evaluation on selected instances to support our claims. We will discuss this at the end of Section 5.4. The general idea behind this is that in a "reasonable" feasible solution, most constraints will trivially be satisfied. Our intuitive approach to constraint generation is the following: Consider two edges that are far apart in the input. If the endpoints of these edges don't move too much while rounding, the edges will also be far apart in the output. Hence, those edges will probably not want to cross, making the sets of constraints preventing them from doing so obsolete. If we could identify and remove all such constraints beforehand, the resulting model would be smaller in size and thus easier to solve. But as the problem at hand is $\mathcal{NP}$-hard, we cannot make good guesses on the relevance of individual constraints.

In general, the opposite is done to implement constraint generation: Creating and solving a possibly underconstrained *partial* model, an external oracle is used to verify correctness of the obtained solution. If the solution is correct, it is also optimal and feasible for the *full* model (containing all possibly relevant constraints); otherwise, the oracle reports back any parts of the solution that would have made it be infeasible for the full model. Then the violated constraints of the full model are added to the partial model. This process is repeated until either a feasible solution is found or all constraints of the full model are added. In either case, the solution produced by the iteratively refined partial model is of the same quality as that of the full model.

We use delayed constraint generation to iteratively add almost all constraints only when needed. We extend the basic model to include the constraints relevant for the consistency of assignment of the sets of binary variables containing all the γ (Equation (5.4)), all the α (Equation (5.6)), and all the β (Equation (5.8)) to get the *empty model*. It is empty in the sense that it is missing all constraints that require a "meaningful" assignment of binary variables; hence the consistency constraints of the empty model can be satisfied almost trivially. In the following, we will use the empty model as the initial partial model to start adding constraints to. Most basic ILP-solvers will pick one decision variable of the empty model at a time and branch on the possible values it can take. For our coordinates this process will implicitly first try moving each vertex to one of the corners of the cell containing it. In the next chapter, in Section 6.2, we reconstruct this procedure when describing the *Greedy Rounding*[4] heuristic; there we also provide an efficient implementation in C++ and discuss the overall performance of this heuristic.

Checking the output of a partial model for topological equivalence can easily be done by any algorithm to test for planarity. To serve as an oracle, we store the original cyclic orders of the input graph and extend the planarity test to also test for consistency with the stored rotation system and make it report back any new incidences, crossing edges, and changed cyclic orders. We then use the results to add the constraints of Equations (5.2), (5.5), (5.7), and (5.9) respectively as needed. The extended partial model is then solved again. Notice that the actual runtime of a planarity test is small compared to the time required to set up and solve the model. Also notice that solving the almost empty basic model takes practically no time. Hence, that way we can easily and quickly find a reasonable initial set of relevant constraints. We support these runtime- and performance claims by observations made in the next section.

5.4 Experimental Performance Evaluation

In this section, we discuss the performance and limitations of the model described above. To do so, we made an implementation using Java to create the model and control the IBM CPLEX solver. We tested this implementation on six graphs of different sizes and complexity. The test instances are hand-picked to convey the impact of vertex count, size of bounding rectangle and number of "difficult" parts on the overall performance.

In the following, the column "*Full model*" is used for executions of the above ILP without any constraint generation. The column "*first*" gives the time until any feasible integer solution (not necessarily optimal) is reported by the integer solver. For both variants, the "*opt*" column gives the time until the solver reports an optimal solution.

We performed the following experiments on a machine with 16 cores (2666 MHz and 4 MB cache each), 16 GB memory and 20 GB swap space and using CPLEX with Java bindings. To compare the full model and the constraint generation approach, we consider model size and wall-clock time spend solving. The measure for model size are the number of rows and columns after CPLEX completes any preprocessing steps; an entry of "†" in a table means that either no model could be created within given time and/or system memory, or that the solver did not report a result within 10 minutes.

For the output figures, white vertices represent the initial positions with the red arrows indicating actual vertex movement. For all input/output drawing pairs, the underlying grid represents the bounding rectangle that was allotted to each instance respectively. As a general measure for complexity of an instance, we will consider the *vertex density* γ, the ratio between the number of vertices in the input and the total number of grid points contained in the allotted bounding rectangle; $\gamma = 100\%$ implies that in the output, all grid points will be occupied.

---

[4] Greedy Rounding will be implemented to obey the constraints of Equation (5.2), whereas the empty model does not.

Small Examples.   We start the discussion with a group of three instances. Each instance has a rather small number of vertices as well as bounding rectangle area. The input drawings are shown in the upper row of Figure 5.7; see Table 5.1 for data on instance size, model size, and time spend solving.

Consider Graph 1 (Figure 5.7 left): The right of the two inner faces contains two additional vertices but the initial drawing has only one available grid point inside that face, making an "unlucky" execution of Greedy Rounding[5] fail to round either vertex. To overcome this, several of these vertices need to make locally non-optimal movements, enlarging the face. Graph 1 is of average *vertex-density* (γ = 36%) compared to the other five instances. Because of its small size and low number of edges, the resulting model is small and thus an optimal solution is found rather quickly. However, its vertices are positioned so that many constraints are not trivially satisfied and significant effort is required even by the constraint generation approach.

Graph 2 (Figure 5.7 center) has lower vertex-density ($\gamma \approx$ 30.5%). In addition, it is designed such that every vertex has one preferred integer grid point that is not contested by any other vertex; hence, greedily rounding each vertex yields the optimal solution. The embedding preservation constraint of the central vertex involves all other vertices, and thus requires most of the α variables to be set properly, whereas in the basic model, these variables can be assigned freely. Any other constraint is easily satisfiable. In terms of computation time, there is not a big difference between finding the first solution and closing the integrality gap. By construction of the input graph, the embedding is trivially preserved by Greedy Rounding; this is also reflected considering model size and solving time of the constraint generation approach. Hence the optimal solution is found by the first run of the constraint generation approach almost immediately. (Note that the 0.5 second runtime includes setting up the Java environment, calling the CPLEX solver and checking topology.)

The input drawing of Graph 3 (Figure 5.7 right) is comparable to those suggested to be input for the area-minimal drawing variant described in Lemma 5.6 – that is, all vertices are initially drawn within the same grid cell. With vertex-density $\gamma \approx$ 77.8%, this is also the most dense instance we did experiments on. First of all, small bounding box result in small and easy-to-solve full models. The size of the bounding box has extreme effect on the runtime – compare the times for the full models of Graph 3 and Graph 1, which has only two more vertices but a much larger bounding box. For such extreme instances, almost all constraints are required and repeated testing and adding of constraints results in an accumulated runtime that is higher than setting up and solving the full model.

---

[5] We will later describe Greedy Rounding to try each vertex in random order, moving it to the cheapest grid point that is available at that time.

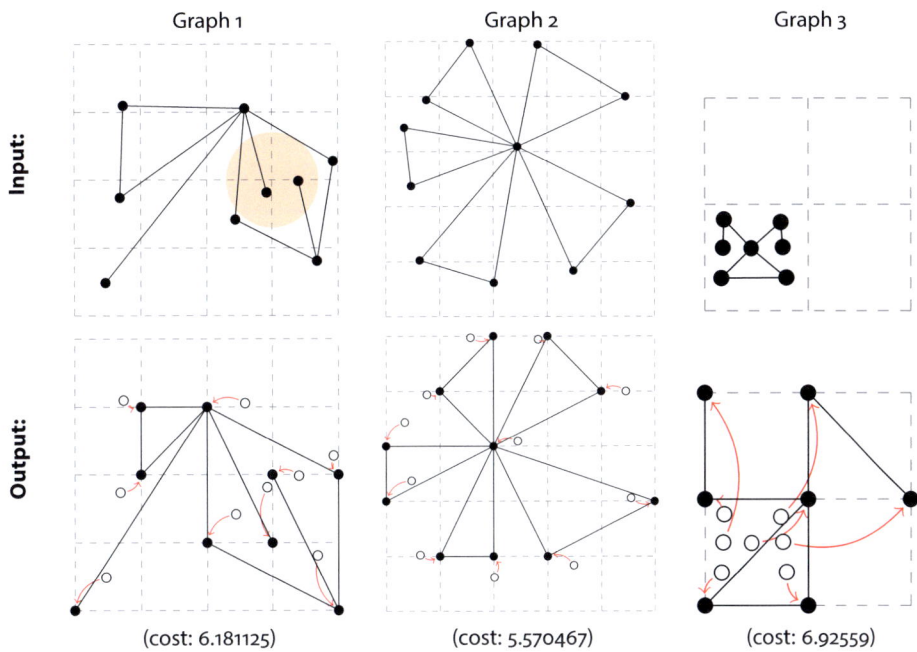

Input:

Graph 1   Graph 2   Graph 3

Output:

(cost: 6.181125)   (cost: 5.570467)   (cost: 6.92559)

Figure 5.7: Graph 1 has one contested grid point (marked by the orange circle) that required significant effort to be resolved. Graph 2 is roundable greedily without complications. Graph 3 is similar to the instances suggested for finding area-minimal drawings, and is supposed to be challenging in general.

Table 5.1: Instance sizes and runtime measurements for Graph 1, 2, and 3, all shown in Figure 5.7.

#	Instance			Full model				Constraint generation		
	$|V|$	$|E|$	γ	rows	cols	first	opt	rows	cols	opt
1	9	10	36.0%	8046	2239	3.2 s	90.6 s	3791	1053	29.2 s
2	11	15	30.6%	26151	6857	5.2 s	10.6 s	2	3	0.5 s
3	7	7	77.8%	2583	896	0.5 s	4.8 s	2245	682	20.2 s

The last example implies that constraint generation is best to be used when the instance has low-vertex density or when many vertices have uncontested preferred grid points to be rounded to. To investigate on these two implications and their relationship, we now look into larger and gradually more complex instances.

Medium-sized Examples. The performance difference between the full model and the constraint generation approach becomes more apparent when the size of the bounding rectangle increases. Consider the two instances of Figure 5.8: They are of about the same density as the first two instances, but double in size; exact measurements on instance size and solving times can be found in Table 5.2.

Graph 4 (Figure 5.8 left) is a path, and thus the model of this instance does not need to contain any constraints to preserve the cyclic orders of vertices. The vertices of the input drawing are spread rather evenly over the bounding rectangle, making almost all vertices have an uncontested preferred grid point to round to. Graph 5 (Figure 5.8 right) on the other hand has one vertex more but is more connected. Graph 5 is also designed to make the Greedy Rounding fail: Trivially rounding the two upper degree-1 vertices will change their relative order around their common neighbor; the path on the right side has several contested grid points and the vertices on it will create new incidences when rounded greedily.

These differences are also represented in the sizes of the full model for both instances. The model for Graph 5 has more than four times as big, mostly because more edges need to be considered with respect to more possible directions. Therefore the solver did not find an optimal solution for the instance of Graph 5 within 10 minutes. To capture the importance of pre-eliminating trivially satisfied constraints consider the final model sizes for the column generation approach on these instances. The lower left part of Graph 5 is a grid on nine vertices, all of which can be greedily rounded correctly. All embedding- and planarity-preserving constraints for edges incident to these vertices are trivially satisfied by any good solution and thus were never added to the partial model. This results in a significant time reduction for solving the partial model to optimality – compare that time to the time it took finding any feasible solution for the full model.

Considering the partial model for Graph 4, we observe that easy instances also benefit from constraint generation, but the obtained speed-up is not nearly as high, even on a smaller bounding rectangle.

Notice that in our context, medium-sized instances only have about 20 vertices and even then, we have to wait for four minutes (in the case of Graph 5) – for most scenarios, graphs of that size would be considered rather small. To emphasize the impact that constraint generation has on instances with large bounding rectangles and low vertex-density, we now consider an even more extreme example.

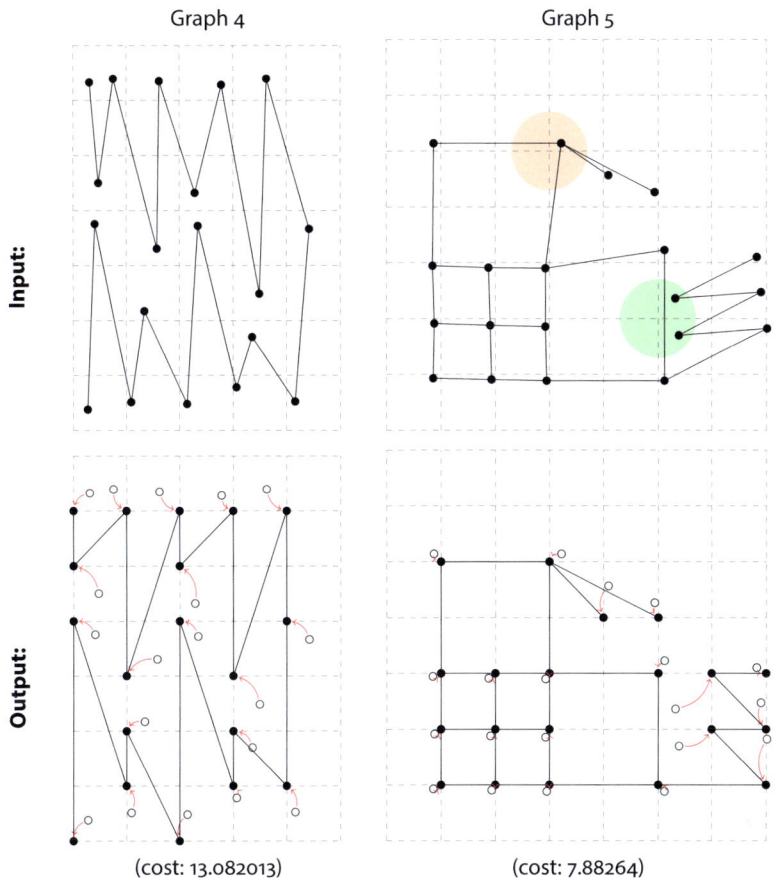

(cost: 13.082013)   (cost: 7.88264)

Figure 5.8: Graph 4 is a path and can be solved optimally fairly easily. Graph 5 is designed to have two complications: One changing cyclic order (orange circle) as well as new incidences and collapsing features (green circle).

Table 5.2: Instance sizes and runtime measurements for Graph 4 and 5, all shown in Figure 5.8.

	Instance			Full model				Constraint generation		
#	$|V|$	$|E|$	γ	rows	cols	first	opt	rows	cols	opt
4	19	18	39.5%	74957	19591	43 s	1106 s	9200	2402	21 s
5	20	25	31.3%	323441	82816	182 s	†	15127	3894	212 s

Input:
Graph 6 (cost: 7.88264)   **Output:**

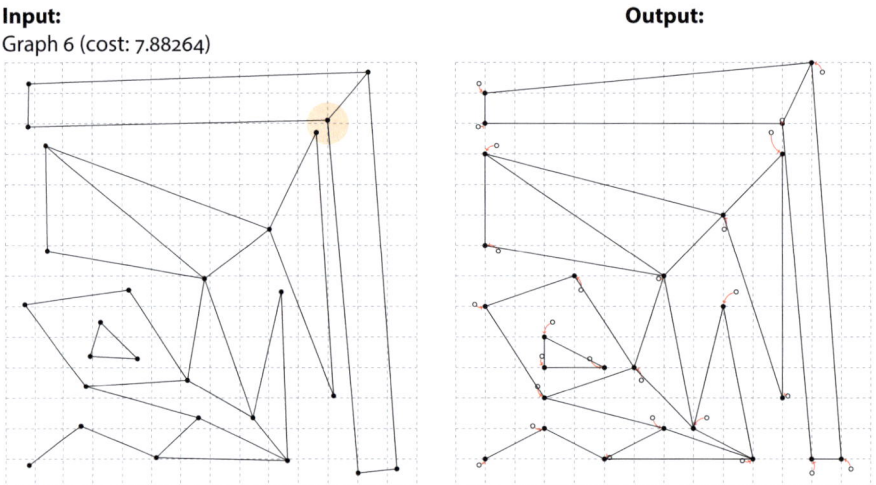

Figure 5.9: Graph 6 is the largest instance we ran experiments on. The only non-trivial part is highlighted by the orange circle at the top-right corner.

Table 5.3: Instance sizes and runtime measurements for Graph 6, all shown in Figure 5.9.

#	Instance $|V|$	$|E|$	γ	Full model rows	cols	first	opt	Constraint generation rows	cols	opt
6	26	34	11.6%	†	†	†	†	12355	3146	7.1 s

A large Example. The last instance we discuss here is shown in Figure 5.9 (with additional data in Table 5.3). In terms of rounding complexity, this instance is almost as easy as Graph 2: there is only one pair of vertices contesting the same grid point in the top right part of the input drawing. Being an easy instance has no impact on the performance of the full model; in fact, creating all possible constraints for all possible directions in this instance already exceeded the allotted computation time and no time was left to spend actually solving the model. In contrast, the constraint generation only had to add one constraint for two vertices of the upper-right corner to the partial model. Rebuilding the model and solving with this constraint runs in reasonable time (compared to the full model). Notice that this constraint does not involve the direction set $\mathcal{D}$.

The key messages to take from this section can be summarized as follows: Small bounding rectangles result in small and quick-to-solve models, but as the bounding rectangle grows in size, so does the model – recall that the number of possible directions $|\mathcal{D}|$ is quadratic in the rectangle's larger dimension and that the number of most constraints is directly linked to $|\mathcal{D}|$. Second, when many constraints are violated during the constraint generation processes, iteratively adding the constraints results in runtime exceeding the time for solving the full model. On instances with moderate vertex-density ($\gamma < 40\%$)

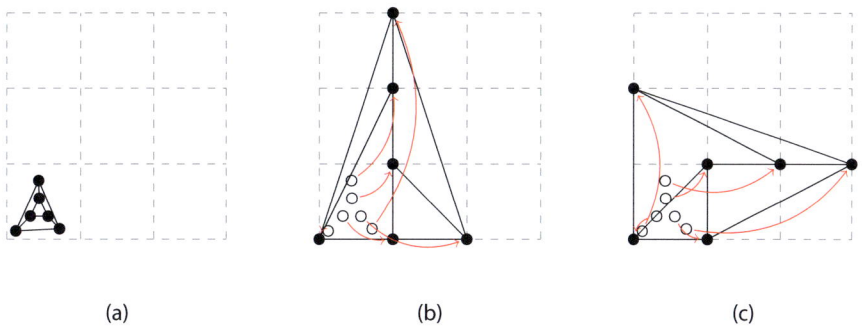

Figure 5.10: (a) A nested-triangles graph on $n = 6$ vertices, placed completely in one grid cell; (b) the output using the rounding objective (Equation (5.1)) – solved in 3 h 25 min; (c) the output using the objective function of Equation (5.10) with $X_{max} = 3$ – solved in 10 min 31 s.

the constraint generation approach clearly outperforms the full model (while still being infeasibly slow in practice).

Graph Drawing. We conclude this chapter with a brief overview on the capabilities of our integer linear program as a graph drawing tool. Area-minimal drawings of planar graphs are useful tools for creating (counter-)examples and thus have a long history in graph drawing, summed up by Frati and Patrignani [FP07]. While Krug and Wagner [KW07] did provide a $\mathcal{NP}$-hardness proof, we do not know of any tool for computing optimal solutions. We demonstrate the capability of our ILP finding such drawings and compare the model tailored to solve TOPOLOGICALLY-SAFE GRID REPRESENTATION to the modified version described in Lemma 5.6. Our model does not contain constraints for fixing a particular outer face; any drawing respecting the given cyclic orders is valid. In Figure 5.10, we show a nested-triangles graph on six vertices (a), together with two area-minimal drawings created using the regular program (b) and the modified version (c). Notice how both drawings have a non-triangular outer face different from the input and that both occupy the same area of six grid cells. Figure 5.10 (b) was created in almost three and a half hours, whereas Figure 5.10 (c) only took about ten minutes. This difference can be attributed to the fact that the objective function of the area-minimizing model takes discrete values (namely the highest y-coordinate), which allows the solver to prove lower bounds faster.

Our model is capable of taking non-planar drawings of planar graphs as input – the lines separating nonincident edges in the input is not represented in the constraints – as well as (in theory) taking non-planar graphs and reporting "infeasible" for them. The latter is strongly inadvisable. We tried solving the model using a drawing of K_5 as input, but canceled computation after twelve hours.

5.5   Conclusion

In this chapter, we have discussed the problem TOPOLOGICALLY-SAFE GRID REPRESEN-
TATION – transforming a given planar drawing into a drawing with the following constraints: Each vertex has coordinates at integer precision, the topology of the original
drawing is preserved, and the total displacement of all vertices is minimal among all
valid transformed drawings.

We have shown TOPOLOGICALLY-SAFE GRID REPRESENTATION to be $\mathcal{NP}$-hard by
giving a reduction from PLANAR MONOTONE 3SAT. To find optimal solutions, we modelled the problem using integer linear programming. We created some selected test instances to evaluate the performance of our original model as well as our faster delayed
constraint generation approach. To do so, we implemented the model using Java and
the IBM CPLEX solver, performing experiments on a virtual machine with 16 cores. We
concluded the discussion of the experiments with empirically analyzing the selected instances, pointing out why they are challenging for our implementation.

Finally, we discussed how our model can be adapted to solve another $\mathcal{NP}$-hard problem, namely minimizing the area of planar straight-line drawings. While applicable in
theory, our experiments have shown the runtime to be infeasible from a practical standpoint.

The results we have presented in this chapter directly motivate those of the following
Chapter 6. The integer linear program discussed here is too slow to be used in any real
world application, hence we next introduce an efficient randomized heuristic based on
simulated annealing.

Chapter 6

Practical Topologically-Safe Rounding of Geographic Networks

In this chapter we consider the Topologically-Safe Grid Representation problem in an application-oriented setting. We are given a geographic network representable by line-segments in the plane – encoded as a planar graph with real-valued vertex coordinates – and consider the problem of representing the vertices at given (integer) grid positions. There are several advantages to such representations as opposed to the common practice of using floating-point numbers for coordinates. Goldberg [Gol91] stated that using integer precision makes explicit what the actual precision of the representation is (because of using a data type), in a data type without mathematical surprises. Milenkovic [Mil95] demonstrated that the original coordinate precision has considerable impact on the precision required to safely perform geometric operations, such as intersecting or calculating overlap. Finding representations on small grids is a natural form of data compression since it reduces the range of the coordinate values. In geographic applications, usually large amounts of data need to be stored and processed. In mobile route planing, for example, the devices in use are often hand-held and have limited resources: small memory, small screens, or slow CPUs. It is also of mathematical interest to consider the smallest grid on which the network can be represented under certain quality constraints. Grid drawings also provide a form of schematization by enforcing a minimum length on edges and introducing a rigid structure in dense areas. This also serves a perceptual purpose: It avoids having arbitrarily small features that need to be drawn. Such features would be hard to read and, ultimately, any visual reproduction of the network is likely to have precision limited to some discrete level anyway.

We are of course interested in *good* grid representations of the input network – for some measure of quality – and not just an arbitrary representation. Many "rounding" and "snapping" procedures from the literature give a bound on the geometric difference between the input and output networks; usually, vertices are allowed to move only within one grid cell. These procedures achieve this by accepting possible changes to the topology of the network, such as allowing vertices to coincide, new intersections to occur, or faces to collapse. We approach the problem from another direction by demanding topological equivalence between the input and output drawings and optimizing the quality of the result. This perspective is motivated by geographic networks, where connectivity, em-

---

A preliminary version of the contents of this chapter has appeared in the proceedings of ACM SIGSPA-TIAL 2019 [vDL19]. This is joint work with Thomas C. van Dijk.

bedding (road networks), and the faces (territorial maps) are crucial. Our main measure of quality for the rounded output is the same as for TOPOLOGICALLY-SAFE GRID REPRESENTATION: the sum of (Euclidean) input-output-position distances for all vertices.

This topologically-safe "rounding" problem is nontrivial, especially if the network has areas where the density of the points is high relative to the size of the grid. In fact, several variants are known to be $\mathcal{NP}$-hard – see Milenkovic and Nackman [MN90], as well as Chapter 5 and our joined work with van Dijk and Wolff [LvDW16] – and no practical method for obtaining high-quality results is known. In this paper we present a practical method based on simulated annealing.

6.1   Related Work and Contribution

Related work on drawing and rounding graphs on the grid can be found in Section 5.1. Here we focus on work about network compression. Recall that the approaches discussed in Section 5.1 bound the distance between input and output, but allow features to collapse. As argued before, these techniques may not be appropriate for geographic applications. We showed in Chapter 5 that minimizing distortion in topologically-safe grid representations is $\mathcal{NP}$-hard in many settings.

As a data compression problem, it is hard to find a minimum representation of arrangements of polygons [MN90]; also recall that the reduction we give in Section 5.2 asks for saving only a single bit on coordinate representations of embedded planar graphs. Shekhar et al. [SHDZ02] give a clustering-based approach to compress vector (road) maps; Khot et al. [KHN+14] present a survey on road network compression techniques.

Contribution.   In this chapter, we consider a variant of the TOPOLOGICALLY-SAFE GRID REPRESENTATION problem from Chapter 5 that does not restrict the output drawing to be contained inside a bounding rectangle. Notice that the hardness proofs from Section 5.2 trivially extend to this less constrained variant. In the following, we use the Euclidean distance – denoted by $\|\cdot\|$ – to measure the cost of vertex movements.

Considering the hardness of this problem, we propose a heuristic approach – in particular, a two-stage algorithm based on simulated annealing. Stage One focuses on finding a topologically equivalent grid representation that we can improve in Stage Two. This is necessary since finding *any* feasible solution that does not, in the worst case, massively distort the input is nontrivial. The second stage uses simulated annealing to improve the quality of the drawing; see Figure 6.1 for two examples on real geographic networks.

The rest of this chapter is organized as follows: In Section 6.2, we establish the basic terminology and introduce easy heuristics that partially solve the problem. In Section 6.3, we give an overview on simulated annealing in general and describe our two-stage algorithm, including pre- and postprocessing procedures. Finally, Section 6.4 describes the setup used for our experiments on real-world networks as well as on artificially generated networks as well as the results. We statistically evaluate the impact of various design decisions on the performance of our approach.

(a)

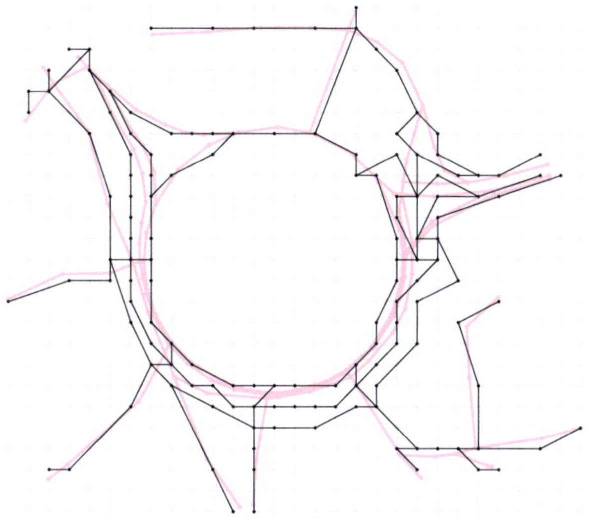

(b)

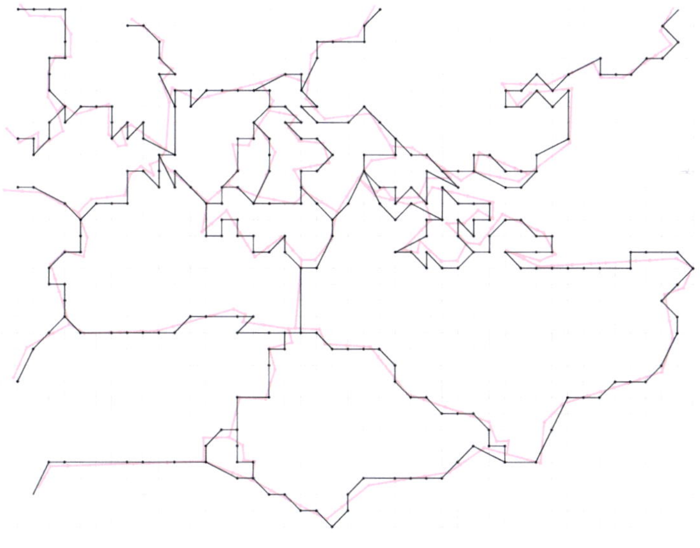

Figure 6.1: Two instances with real-valued input (light drawing) and corresponding grid representation (black drawing) computed using our algorithm. (a) Roundabout in downtown Würzburg (138
vertices, 155 edges), grid size 28×28, average vertex movement 0.600, computed in $15\,s$; (b) borders in
Britain (3110 vertices, 3207 edges), grid size 240×240, average vertex movement 0.435, computed in
$70\,s$, cropped to show only the south-east (roughly centered around London).

6.2   Terminology and Basic Heuristics

Given an input drawing Γ_1, our two-stage approach will produce *intermediate* drawings Γ_k – drawings where some vertices are already moved to the integer grid while others are not. We call a vertex *nongrid* if its coordinates are not integer. As before, the *cost* of a vertex v in a drawing Γ_k (with $k > 1$) of G is defined as the Euclidean distance between its original (real) position $p(v)$ in the input and its position $r(v)$ in the drawing Γ_k; the cost of a drawing is the sum over the costs of its vertices. This matches the notation and objective function stated in Section 5.3.1. A drawing is called *feasible* if it is topologically equivalent[1] to the input and all vertices are positioned on grid points. Note that the original drawing Γ_1 always has cost zero, but – except in trivial cases – is infeasible.

In the iterative procedures we describe in this chapter, we proceed from one intermediate drawing to the next by *moving* a vertex: A *move* is the change of position of a single vertex of the current drawing, that also keeps the position of each other vertex unchanged. A move is called *valid* if it does not change the topology of the drawing, and the *cost* of the move is the difference in cost between before and after the move.

As a baseline, we describe several rather straightforward *partial* heuristics. They relate to *rounding* vertices as known from the literature, as they move vertices to the closest available grid point; they are partial in the sense that they are not guaranteed to find a feasible solution.

Instant Rounding.   Independently set the position $r(v)$ of each vertex to the nearest grid point. If the resulting drawing is topologically equivalent to the input, it is an optimal solution since every vertex independently moves the minimum amount possible. Otherwise the heuristic failed, returning the original drawing unchanged.

Incremental Rounding.   Round the vertices one by one, in arbitrary order, checking the validity of each move. Undo any invalid moves, leaving each such vertex v in their original position $p(v)$. This procedure may succeed in rounding a subset of the vertices even if it fails to solve the entire instance; this can be useful for quickly solving "easy" parts of the graph. The cost of this drawing is still a lower bound on the optimum since any vertex that moves does so by the minimum amount.

Incremental Greedy.   Do the following for all vertices, one by one, in arbitrary order: consider moving it to any of the four corners of the grid cell it is in.[2] Check these candidate moves for validity in nondecreasing order of cost (distance to $p(v)$); accept the first valid move and continue with the next vertex. Vertices with no valid candidates remain unrounded. Note that the rounding cost of this (partial) solution is not a lower bound on the optimum, even if it finds a valid grid position for all vertices since greedy decisions were made.

---

[1]   Recall that *equivalence* is stronger than the *topological similarity* stated in the definition of snap rounding (Definition 5.1 (3) on page 61).

[2]   This is ill-defined for vertices on grid lines; in fact we take the floor or ceiling on each axis, resulting in *at most* 4 candidate positions per vertex. Hence, vertices that already lie on a grid point are not moved.

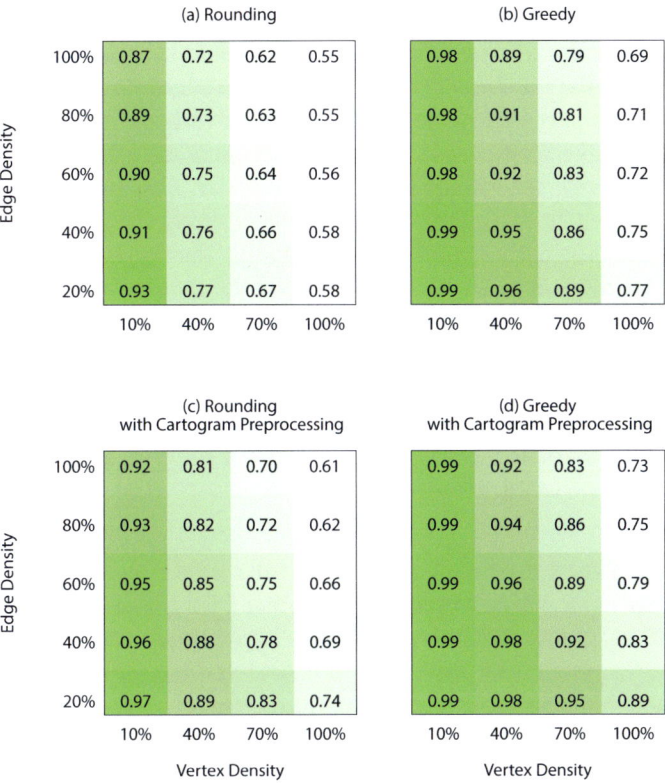

Figure 6.2: Success rates of (a) incremental rounding, and (b) incremental greedy as described in Section 6.2. Figures (c) and (d) include the cartogram preprocessing from Section 6.3.3. Each entry is for 100 random instances with area $[0, 19] \times [0, 19]$ and varying vertex and edge density (see Section 6.4.1 for a description of the random instances). Matrices are sparsified for readability.

Depending on the properties of the considered instances, these easy heuristics may work quite well or experience significant difficulties. For example, in Figure 6.2 (a) and (b), success rates of these heuristics are given for randomly generated planar graphs of various vertex density γ and edge density ε on a fixed area. These instances and how they are created is described in more detail in Section 6.4.1; here we include vertex densities from 10% up to 100% (meaning one vertex per grid point on average), and edge densities from approximately matching-sized edge sets (20%) up to a full Delaunay triangulation (100%).

The matrices show the success rate of the Incremental Rounding and Incremental Greedy algorithms on 100 graphs for each setting; that is, numbers and colors indicate the fraction of *vertices* in the output drawings that are at integer-precision coordinates.

We see that the greedy approach clearly performs better than simple rounding, but even in the sparsest graphs, rounding fails to move 1% of the vertices to the grid on average. The gradient of the values indicates that vertex density seems to be more challenging for both approaches than the number of edges: although the edges also constrain the feasibility of drawings, it is particularly the (local) density of vertices that forms a problem for these heuristics. Note, for example, that a grid cell containing more than four input vertices cannot be completely rounded by these heuristics; this, among other difficult situations, is more likely to occur in instances of higher vertex density. Scale & Greedy (described below) overcomes this problem by exploding the size of the network, effectively increasing the number of available grid points at the expense of a large (rounding) cost.[3] This is efficient in terms of runtime, but often requires large scaling factors on real data and therefore returns drawings that are completely unacceptable as a heuristic and impractical for Stage Two. We support this claim by experimental results, see Section 6.4.2.

We also present a heuristic that always finds a feasible solution. However, it can give very bad solutions, as we will discuss in Section 6.4.2.

Scale & Greedy.   Repeat the following steps in order until a valid drawing is found. Uniformly scale the input network by a factor f (initially $f = 1$) and apply the Incremental Greedy algorithm. If this results in no vertices being nongrid, we are done; otherwise, increase f by one and retry. This process terminates since Incremental Greedy will succeed for large enough f. However, scaling the entire network is likely to result in high cost.

6.3   The Two-Stage Algorithm

We do not allow topological alterations of the network and insist on moving all vertices to the grid, thus we have to handle vertex-dense clusters. To do this, we first propose and discuss several approaches. The difficulty of TOPOLOGICALLY-SAFE GRID REPRESENTATION lies not only in finding a solution with low cost since it is already nontrivial to find any reasonably similar feasible solution at all. Given these difficulties (and the intractability of finding optimal solutions), we now consider a local search approach. Since we do not want to require starting with a feasible solution, this local search will have to handle infeasible states: it needs to consider drawings in which there are (still) nongrid vertices. However, it will never consider drawings that are not topologically equivalent to the input: Topological equivalence of the current drawing and the input drawing is an invariant we maintain at all times, rather than a property we are searching for. All our search algorithms use the following definition of transitions between neighbors for intermediate drawings:

---

[3] From a data compression perspective, scaling by a factor two corresponds to using an additional bit on each axis of each vertex. With enough bits, the drawing can be represented without significant rounding.

Local search neighborhood transition.   Take any vertex v and perform any valid move among the following. If v is nongrid, move it to a random corner of the grid cell it is currently in; otherwise, move it to one of the eight grid points surrounding it.

Two drawings are neighbors in the local search, when they differ only in the position of a single vertex and when this difference is obtainable by the transition defined above.

If we use *hill climbing* (always greedily picking the neighbor with lowest cost) in this neighborhood definition, nothing happens: the state representing the input drawing Γ_1 already has cost zero and any neighbor of Γ_1 is more expensive. Since we generally foresee further complications due to local optima, we pick the well-known metaheuristic *simulated annealing* by Kirkpatrick, Gelatt, and Vecchi [KGV83]. For a description of simulated annealing, see for example Van Laarhoven and Aarts [vLA87]; below we briefly sketch the general approach. Simulated Annealing borrows its terminology from metalworking: We will use terms like energy, temperature and cooling, which we define next.

Simulated Annealing.   Simulated annealing is an iterative local search procedure. We consider all topologically valid drawings of the input network as possible *system states*. We define the *energy* $E(\Gamma)$ – or *cost* in our terminology – of a state Γ to be the rounding cost as defined in the problem statement of TOPOLOGICALLY-SAFE GRID REPRESENTATION. To transition from state Γ_k to Γ_{k+1}, we pick a random (valid) move from the *neighborhood* described above. If the new state has less energy than the current one $(E(\Gamma_{k+1}) < E(\Gamma_k))$, we *accept* it as our new current state. If the new state has the same or more energy – that is, it is worse – we can still accept it, and do so with probability $\exp(-\delta/T)$, where δ is the difference in cost and T is a variable called the *temperature*. This randomness allows simulated annealing to escape local optima. For the *cooling schedule*, we employ the standard exponential schedule $T_n = c \cdot T_{n-1}$ for some constant factor c; in this way, it becomes less likely to accept worse solutions as the search progresses. We discuss the impact of different values of c in Section 6.4.3.

We experimentally observe that a straightforward annealing approach using cost as the objective function does not perform well (for an example, peak ahead to Figure 6.4 (a) on page 101): the focus on cost often prevents it from finding a feasible drawing. One possibility to overcome this would be to design an objective function that rewards feasibility. However, this imposes several difficulties. First, in order for the search procedure to actually *find* the feasible drawings, this added term must be "smooth" enough to provide attraction, but it is not clear how to do this. Furthermore, depending on the details of this added term, it would have to be tweaked to not interfere with the cost optimization too much. We are eager to avoid such extra tuning parameters, which would come on top of the parameter tuning required for the simulated annealing itself. Therefore, we take a different approach by splitting the algorithm into two stages: One focused on finding a reasonable feasible drawing quickly, and a second straightforward annealing phase tries to minimize the total cost of all vertex movements. We conclude the section with optional preprocessing and postprocessing steps, and some algorithmic implementation considerations.

6.3.1 Stage One – Feasibility

We describe several feasibility procedures. Their goal is not to minimize the total move-ment cost, but to efficiently find a feasible drawing that can then be optimized for cost. We first sketch an approach that is guaranteed to find a feasible drawing quickly. Unfor-tunately, these drawings have impractically high cost and do not serve well for Stage Two. Then we describe a local search procedure that works well in practice.

Graph drawing. Take the input as an abstract (embedded) graph and draw it anew, ignoring the vertex positions given in the input drawing. This can be done, for example, with the algorithm of Harel and Sardas [HS98] which deterministically computes a com-pact grid drawing preserving a given embedding in linear time. Besides several technical challenges, such as requiring the graph to be biconnected, our experiments indicate that these drawings do not provide a good starting point for our Stage Two: they are too dis-similar to the input in terms of shape and positions. See Section 6.4.2 for a qualitative evaluation of this approach.

Vertex-density annealing. This procedure uses the local-search neighborhood in-troduced above, but with a different objective function. Since locally dense regions are hard to successfully round, we want the search algorithm to make space. We therefore minimize the sum of inverse squared distances for all pairs of vertices.

$$f_{\text{density}}(p) = \sum_{v,w \in V, v \neq w} \frac{1}{\|p(v) - p(w)\|^2} \tag{6.1}$$

We now use simulated annealing to reduce this score by moving vertices away from each other in a topologically safe way one step at a time – either putting it to one of the four nearest grid points (cell boundaries) or moving to one of the neighboring eight grid-points. This function is modeled to be reminiscent of the repulsive forces in force directed graph drawing (for further details, see Eades [Ead84]). Notice that minimizing this function is easy: Simply scale the graph to be arbitrarily large. Hence we do not *actually* want to minimize this objective function but rather run the search at a constant temperature of $T = 1$ and terminate the search as soon as the drawing is feasible.[4] We unconditionally accept any valid move that moves a nongrid vertex onto a grid point, regardless of its effect on the cost since that is the real goal of this stage. Constantly keeping the temperature at a relatively high level ensures enough freedom of movement: The goal of this stage is to escape locally difficult situations.

Here are some observations about Vertex-density annealing. As we consider squared distances, a cluster of very close vertices will have strong impact on the energy of an otherwise sparse network. If the score of a vertex is low, it is more likely that greedy snapping will be safe as small changes can only cause problems when grid points are

---

[4] This makes "annealing" a bit of a misnomer, but for uniformity of presentation, and since we do have Kirkpatrick-style move acceptance, we call it annealing.

contested. Every network always has a density-reducing move, as some vertex on the outer face can always move away from the others. Making small moves keeps vertices rather close to their input coordinates.

Grid-density annealing. This works identically to Vertex-density annealing, except it interprets the density more locally based on the grid using the following procedure. Every nongrid vertex adds $\frac{1}{4}$ "density" to its four surrounding grid points and every other vertex adds $\frac{1}{9}$ to its eight surrounding grid points and the one it is placed on. That way, vertices effectively charge the grid points they could possibly move to within one iteration; grid points with high values are more likely to be contested by multiple vertices. Then we say the score of a vertex is the squared density of the grid point it is on – or, for a nongrid vertex, the squared density of the nearest grid point. The annealing objective value is the sum of these vertex scores. As with Vertex-density annealing, this encourages vertices to move out of the way of other vertices, and particularly provide space for nongrid vertices (since they contribute more to the density). However, it is not immediately clear if the more local nature of Grid-density annealing is good or bad. This is evaluated in Section 6.4.2.

Improvements based on Structural Considerations. In order to find a feasible drawing sooner, we augment the neighbor selection by adding two improvements that are both based on the particular structure of the problem at hand. Both further depart from the classic simulated annealing approach described above by changing how the local neighborhood of a given drawing is generated and explored.

First, at each step, we perform the Incremental Greedy algorithm in addition to the regular local search move, immediately moving any nongrid vertices to the best available grid point on its cell. This shortcuts having to wait for the random vertex selection to pick such vertices eventually.

Second, rather than selecting a vertex uniformly at random, we temper how vertices are sampled in two ways. We select the vertex according to a distribution based on its individual contribution to the total density measure of the current state. This should encourage the algorithm to search in dense areas, hopefully giving progress toward feasibility. Additionally since we try moving all nongrid vertices before creating the next neighboring drawing in every iteration, we only try moving vertices already on the grid – those are the vertices that block grid points, possibly blocking other vertices from being put onto the grid. See Section 6.4.2 for a statistical evaluation discussing these modifications.

6.3.2   Stage Two – Reducing Cost

Once Stage One finishes by finding a feasible drawing, we switch to straightforward simulated annealing with the objective function we originally intended – minimizing rounding cost. Since this is a more traditional annealing approach where we want to avoid local optima, we pick the typical exponential cooling schedule. See Section 6.4.3 for details of the parameter selection. Using any Stage One strategy will move vertices away from their original position – some more, some less, but the network is likely to expand. Stage Two tries to undo the expansion while maintaining topology and coordinate precision – staying on the underlying grid.

To transition between two neighboring states Γ_k and Γ_{k+1}, we randomly pick a vertex $v \in V$ and randomly mutate its current position by independently mutating its (integer) coordinates – adding or subtracting 1 from x_v and/or y_v. Given that v was on the integer grid before, it will be on the integer grid afterwards – the mutation will move v to one of the eight (octilinear) neighboring grid points.

The way we generate these drawings does not check topological equivalence – rotation systems might change, vertices might end up on other structures or edges might cross. To overcome this, we rely on a modified acceptance function: after generating some neighbor Γ_{k+1}, we check it for topological consistency and immediately reject any inconsistent drawing. If Γ_{k+1} was not rejected, we evaluate the energy level $E(\Gamma_{k+1})$ and follow the original approach proposed by Kirkpatrick et al. [KGV83]: If the energy level (the total movement cost) is lower, we accept Γ_{k+1}; otherwise we can still possibly accept it, even if the energy level is higher. The probability for accepting an energy-increasing move is given in Equation (6.2).

$$\exp\left(-\frac{E(\Gamma_{k+1}) - E(\Gamma_k)}{T}\right) \tag{6.2}$$

6.3.3 Pre- and Postprocessing

To improve the performance of our approach, we suggest a preprocessing procedure and a postprocessing procedure to be performed before and after Stage One and Two respectively. Both procedures are efficient and deterministic; we evaluate the impact of pre- and postprocessing on the overall performance of each stage in Section 6.4.

Preprocessing: Linear Cartograms. Since dense areas of the input drawing are hard to resolve, some preprocessing to assist Stage One seems appropriate. Hence, we propose expanding the input using efficiently computed *linear cartograms* by van Dijk et al. [vDH14, vDL18]. In a linear cartogram, all edges of a drawing of a network are drawn with a prescribed length. Clearly, not every combination of network and prescribed edge lengths is possible.[5] To minimize the error in edge lengths, this method is based on finding a solution to an overconstrained set of linear equations using least squares adjustment. For our problem, we create a twofold set of equations. On one side, we ask for any edges shorter than length $\sqrt{2}$ (the diagonal of a grid cell) to be elongated and for vertices that are too close together to be moved apart; on the other side, we want the vertices to stay relatively close to their input positions. We will see in Section 6.4.2 that using linear cartograms is quite effective in practice.

---

[5] For example, consider a triangle with two short edges and one edge that is longer than both other edges combined.

This preprocessing is implemented using the following linear least-squares adjustment formulation, computing new positions p given the original positions r. Note that we create an overdetermined system of equalities on purpose; least squares adjustment will find an optimal compromise between the conflicting constraints. (See for example Kraus [Kra11] for a general introduction.)

First, in the interest of the cost of the final drawing, vertices should ideally stay where they are. For every $v \in V$, we therefore have the following *vertex position* constraint from Equation (6.3) on the x and on the y axis.

$$p(v) = r(v) \qquad (6.3)$$

The relative position of vertices that are connected by a long edge should also remain the same, implying that the length of that edge stay the same. Edges shorter than $\sqrt{2}$ are problematic since it is likely that both endpoints contest for the same grid position. We want to introduce some additional distance between the two vertices of any short edges, stretching the edge towards length $\sqrt{2}$. Hence, we have one of the two *edge length* constraints for any $(u, v) \in E$ of the input and on both axes – either from Equation (6.4) for long edges or from Equation (6.5) for short edges:

$$p(v) = \begin{cases} p(u) + r(v) - r(u) & \text{if } \|r(u) - r(v)\| > \sqrt{2}, \text{ or} \quad (6.4) \\ p(u) + \dfrac{\sqrt{2} \cdot (r(v) - r(u))}{\|r(v) - r(u)\|} & \text{otherwise.} \quad (6.5) \end{cases}$$

We also add the constraints of Equations (6.4) and (6.5) to the non-edges missing from a constrained Delaunay triangulation. A *Delaunay triangulation* is a triangulation of a set of points in the plane such that the circumcircle of any triangle does not contain other points. A *constrained* Delaunay triangulation departs from the original idea by enforcing that the triangles cover a prescribed set of edges (possibly violating the circumcircle-rule to do so, for more details see Chew [Che93]).[6]

In addition, we put additional emphasis on separating any pairs of nonadjacent vertices that are too close, adding *vertex distance* constraints similar to Equation (6.5). See van Dijk et al. [vDvGH+13] for more on constrained Delaunay triangulations when transforming geographic networks.

Since we want to punish violations of the different sets of constraints differently, we add weight factors to the misclosures[7] of the different equations. Intuition suggested that preventing short edges is most important, whereas trying to keep vertices at their original positions is conflicting all other constraints. Also, we deemed pushing apart any vertices that are too close to be more important than preserving the constrained Delaunay triangulation. Following these assumptions, we have considered a base setting

---

[6] In the case of artificial instances (look ahead to Section 6.4.1), a constrained Delaunay triangulation does not violate the rule – it simply re-adds the previously discarded edges. Constrained Delaunay triangulations of real-world instances are most likely not Delaunay triangulations.

[7] The *misclosure* of an equation is the amount by which left-hand and right-hand side differ.

Table 6.1: Different weights tested for the Cartogram preprocessing.

Parameter	base	low	high	final	Equations
vertex position	0.2	0.0001	1.0	0.8	(6.3)
edge length	4.0	0.1	8.0	4.0	(6.4), (6.5)
constrained Delaunay	2.0	0.1	4.0	2.0	(6.4), (6.5)
vertex distance	1.0	0.1	4.0	1.5	(6.5)

for the weights – see the first column of Table 6.1 – and from these values, we pitched each weight to either extreme individually – as in the second and third column of Table 6.1, respectively –, leaving the other three unmodified. We finally settled for the values shown in the fourth column of Table 6.1. Those will also be the values that we use throughout the rest of this chapter whenever we use cartogram preprocessing.

We only modify vertex positions according to the weighted minimum of misclosures, but do not check for topological consistency. Hence, the resulting drawing $\Gamma(G, p)$ using the new positions might not be topologically equivalent to the input. In that case, van Dijk and Haunert [vDH14] describe a back-off procedure that we use here: Consider linearly interpolating the vertex positions uniformly from in the input drawing to those obtained by the least squares adjustment. The interpolation factor (ranging from 0 to 1) can be seen as a time step. We find the latest discrete time step in this interpolation that yields a valid drawing – by starting at 1 and testing 0.1-decrements – and output the corresponding drawing.

Postprocessing: Hill Climbing. Our simulated annealing algorithm chooses the next state to evaluate at random. Annealing theory suggests that by extending the cooling schedule, the probability for the algorithm to find a global optimum converges towards 1 (see Granville, Krivanek, and Rasson [GKR94]). As we have no efficient means of telling if a solution is optimal or not, we have to stop Stage Two eventually and take the last accepted state as final drawing. Annealing for longer could possibly yield better results, but by our choice of cooling schedule, the annealing temperature goes to zero and the search reduces to local hill climbing. Hence we propose a straightforward hill climbing implementation as a much more efficient postprocessing: rather than sampling random vertices and attempting moves, our approach iteratively applies valid moves that improve the outcome until a local optimum is reached. This suggests the possibility of annealing at a higher temperature than one normally would, and relying on the final hill climb to clean up the solution. See Section 6.5 for an evaluation on the trade-off between hill climbing and annealing for longer time.

6.3.4 Implementation Considerations

Our algorithm often needs to test whether a particular move is valid. The expensive part in this is checking for possibly intersecting edges. Rather than the well-known (and worst-case more efficient) Bentley-Ottmann sweepline algorithm [BO79] for line segment intersections, we implement the following bin-based approach to speed up these tests. First, we overlay a $W \times W$ regular grid of rectangular bins over the full extent of the current drawing, then loop over all edges and put them in all bins that they intersect, and finally check all pairs of edges in each bin by brute force. If W is picked large enough, at most a small number of edges will be in any particular bin. This is efficient on real-world data for a wide range of values W. Our code uses $W = 512$ as a somewhat arbitrary trade-off between memory, number of bins, and the population of bins. For more details on this method, see also Peng and Wolff [PW14].

In fact, this grid-based approach – at least for our application and on our data – outperforms the CGAL implementation [ZWF19] of the Bentley-Ottmann algorithm by more than two orders of magnitude, even though CGAL generally has high-quality and high-performance implementations. Its problem seems to be the numerical instability of the sweepline algorithm, which requires CGAL to use high-precision arithmetic on our real-world data – something our relatively crude algorithm does not require. We additionally point out that our current implementation is single-threaded, but in principle checking the bins can be easily parallelized.

We use the *Computational Geometry Algorithms Library (CGAL)* for basic geometric computations and for constrained Delaunay triangulations (see the CGAL manual [Yvi19] for details). We use the Eigen library [GJ⁺10] for highly-efficient sparse matrix calculations in the cartogram code.

6.4 Experimental Results

In this section, we statistically evaluate various properties of our algorithm and the effectiveness of design decisions and parameter choices. Whenever we directly compare two options, we provide a (two-sided) Wilcoxon paired signed-rank test [Wil45] and report the z-score, demonstrating statistically significant improvements; recall that a z-score of 1.96 or higher satisfies a 95% confidence level.

6.4.1 Test Instances

We provide experiments on both real-world networks and artificial instances. For the real networks, we have used OpenStreetMap shapefiles[8] and the City of Chicago Open Data Portal[9]. Since these road networks are not always planar, we preprocessed the instances, introducing vertices at any intersections in order to get plane graphs.

We also consider random artificial instances. This allows us to perform systematic and statistically significant experiments on large datasets without having to select and sanitize real-world road networks. These instances are based on vertices sampled using binomial point processes, placing v vertices uniformly at random in a square area of X

---

[8] `https://www.geofabrik.de/data/download.html`
[9] `https://data.cityofchicago.org/`

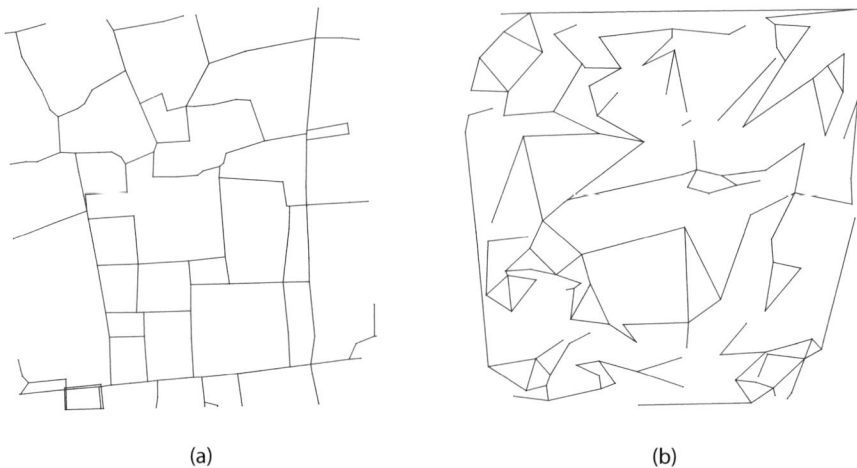

(a)   (b)

Figure 6.3: Visual comparison of real-world and artificial networks: (a) Würzburg downtown with 134 vertices and $\varepsilon = 40.5\%$; (b) an artificial instance with roughly the same parameters.

by X units – thus having $X + 1$ possible integer coordinate values in each dimension. To generate instances with many edges, we take the Delaunay triangulation [GKS92] of this point set; for instances with fewer edges, we take a subset of the triangulation edges as described below. We therefore have three parameters describing size and complexity of an instance: the side length X of the area in which the points are sampled, the number of vertices v, and the *edge density* – the number of edges as a percentage ε of those present in a complete Delaunay triangulation. As an alternative to specifying the number of vertices v, we can also consider the *vertex density* γ. This is the ratio between the number of vertices and the number of grid points of the area; up to rounding we have $v = \gamma \cdot (X+1)^2$. Throughout the rest of the chapter, we describe random instances by the triple of these parameters, namely as (X, γ, ε).

We can also determine γ and ε values for real-world instances by calculating a Delaunay triangulation of the point set. Vertex density can be interpreted as desired compression rate. While testing our algorithm on real-world instances, we noticed that aiming for $\gamma < 40\%$ gives good results in reasonable time.[10] Therefore, we will also focus our experiments on artificial instances on those with vertex density $\gamma = 40\%$.

Edge density is independent of the size of grid cells. We have seen in Figure 6.2 that the impact of higher edge density is less severe than that of higher vertex density when considering how complicated rounding that instance is. The road networks we consider typically have an edge density value ε between 35% and 45% – therefore we focus on artificial instances with $\varepsilon = 40\%$ and $\varepsilon = 100\%$. For comparison, see the instances listed in Table 6.2. When dropping edges from a Delaunay triangulation to obtain lower values

---

[10] Higher vertex densities resulted in runtimes that were generally higher and also of larger variance.

of ε, we need to make sure that we do not create isolated vertices. Our algorithm was designed with road networks in mind: while road networks can easily be disconnected, they usually do not have isolated (unreachable) vertices; hence, our algorithm cannot handle such vertices safely. To avoid this problem in our artificial instances, we proceed as follows. We sample the point set P and calculate a Delaunay triangulation. Then we compute an arbitrary depth-first search tree of this triangulation and take a maximal matching from the tree's edges: if $|P|$ is even, that matching is perfect (as shown by Dillencourt [Dil87]), otherwise we add one extra edge to cover the missing vertex (creating a path of length two); in either case, we mark all edges obtained this way. We then delete a uniformly random set of the other edges to achieve the desired number of edges. Since the marked edges cover every point of P, no vertices end up being isolated in the final instance. A visual comparison between real and artificial instances is provided in Figure 6.3: on the left is a part of Würzburg downtown, on the right is an artificial instance with approximately the same density parameters.

6.4.2   Evaluating Stage One

As discussed before, the basic rounding heuristics will not consistently find feasible drawings and those provided by the graph drawing algorithm are too distorted for the annealing process to find a good solution in reasonable time. In this section, we evaluate the performance of the procedures for Stage One from Section 6.3.1 on artificial instances. As a baseline, we first consider the simplest variant, that is, without greedy steps during the annealing and sampling all vertices uniformly. The experiments in this section are run on 1 000 random instances with configuration $(19, 40\%, 100\%)$, that is, 400 grid points in the sampled area and therefore 160 vertices and approximately 450 edges.

Before deciding between the different advanced procedures, we first rule out Scale & Greedy and Redrawing the embedded graph; then we compare Vertex-density annealing to Grid-density annealing, with and without cartogram preprocessing and other improvements over a traditional annealing approach. To get an intuition regarding the total cost of an instance recall the definition of snap rounding (Definition 5.1 on page 61). Enforcing geometric similarity would imply a maximum cost of $\frac{1}{2}\sqrt{2} \approx 0.71$ per vertex (or about 113 per 160-vertex instance). While we cannot hope for such low costs, our goal for Stage One is to stay within the same order of magnitude.

Scale & Greedy is bad.   We have seen in Section 6.2 that Incremental Greedy is bound to fail if there are too many vertices inside the same grid cell. Scale & Greedy works on the idea that adding more grid lines – and thus more possible coordinate values – will eventually make Incremental Greedy work. Instead of inserting new grid lines – moving from integer to half integer, then to quarter integer and so on – we do the opposite, scaling the drawing (and all coordinates) by a constant factor. By doing so, we preserve the notion of "moving to the integer grid" and report scaling factors instead of additional bits needed for representing the refined grid. We ran the following experiment, searching for the right factor to scale the original coordinates with starting at 1; once a feasible solution is found, we report the factor and the total vertex movement induced by scaling and rounding. The instances in this experiment admit reasonable solutions with a cost of about 500 (as we will see). Scale & Greedy required an average scaling factor of 3.86 (ranging between 2 and 23), resulting in an average cost of 3 329 and no solution with cost below 1 047.[11] This holds true also for real-world instances, e.g. producing a feasible drawing of the roundabout in Würzburg from Figure 6.1 (a) requires a scaling factor of 4. This disqualifies Scale & Greedy as a practical Stage One procedure.

Redrawing is worse.   We created grid drawings of all instances using an implementation of the algorithm of Harel and Sardas [HS98] created by Johannes Zink [CLWZ19]. In the context of graph drawing algorithms, our instances are small and thus grid drawings can be computed quickly compared to our randomized annealing approaches. One could imagine that the time saved in Stage One could be spend optimizing the output in Stage Two for longer. However, the average cost of these drawings is an enormous 26 256 (or average cost per vertex of 164). This makes it is completely impractical – consider for example that any move in Stage Two can improve the cost by at most $\sqrt{2}$.

Vertex density vs. Grid density.   In Section 6.3.1, we have proposed two different "density" objective functions and claimed that immediately optimizing cost leads to difficulties finding feasible a solution. We now experimentally evaluate this.

We ran the three options – cost, Grid-density, and Vertex-density – on each graph for 20 000 iterations. Whenever a feasible solution was found within the allotted number of steps, the remaining iterations were spent in Stage Two by optimizing the output drawing for total movement cost. See Figure 6.4 for the behavior on a typical instance over time (counted by iterations): The purple line shows cost, the turquoise line shows the number of vertices that have taken a grid positions.

A general trend over the 1 000 instances is that Vertex-density annealing generally requires fewer steps to find a feasible solution, whereas Grid-density annealing generally finds a (first) feasible solution with lower cost; simply annealing with the objective of phase two – minimizing rounding cost – fails to find a feasible solution at all on 362 of the 1 000 instances, oftentimes being stuck on the last one or two vertices. Since it is not better than Incremental Greedy, this disqualifies using annealing for rounding cost to find feasible drawings.

On the other hand, a Wilcoxon test shows that the differences between Vertex-density and Grid-density are significant: Vertex-density annealing finishes significantly sooner ($z \approx 28$, "winning" 688 of the 1 000 instances), but with higher cost ($z \approx 10$). As a qualitative indication, when it finished first, the Grid-density solution was 15% cheaper on average (347.208 compared to 407.916). On the instances where Vertex-density was faster, the average cost of the solutions found was higher: 592.480. This suggests that

---

[11] Here we measure cost by aligning the centers of both drawings so that scaling up moves vertices away from the center.

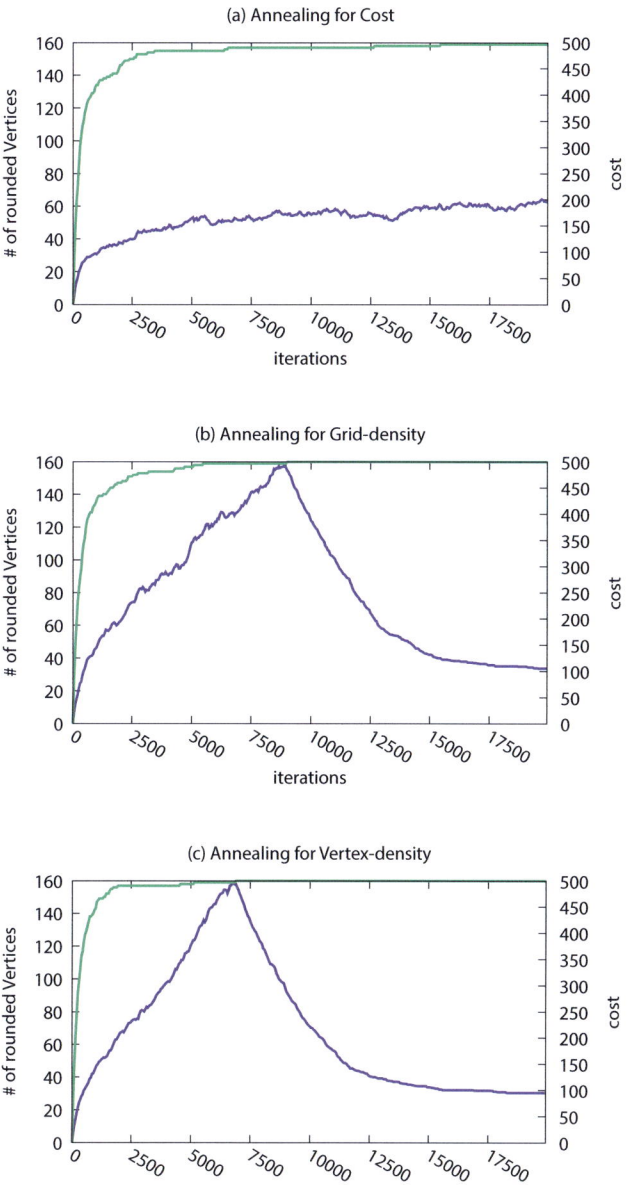

Figure 6.4: Qualitative evaluation of Stage One performance on a single artificial instance with $(19, 40\%, 100\%)$; annealing Stage One for: (a) cost, (b) Grid-density, and (c) Vertex-density. Purple line shows total rounding cost, turquoise line shows the number of rounded vertices.

Grid-density is quicker at finding relatively easy solutions, but struggles on more difficult instances.

Cartogram Preprocessing.   The main goal of linear cartograms from Section 6.3.3 is to reduce the time Stage One needs to find a feasible solution as well as to reduce the total rounding cost of the found solution. Concerning the time taken to find a feasible solution, we observe the following: Both procedures got significantly faster – for Vertex-density, we get $z \approx 25.95$, and for Grid-density, we get $z \approx 23.35$ – and while both procedures improve, it still holds true that annealing for Vertex-density is generally the faster option for Stage One ($z \approx 22.47$).

Considering the total cost of the feasible solution, we notice that those found by Vertex-density annealing got cheaper ($z \approx 3.828$), while we failed to show an improvement for the Grid-density annealing ($z \approx 0.124$). Again, the observations we made above are still true, and Grid-density annealing finds significantly cheaper solutions ($z \approx 26.53$).

This demonstrates that preprocessing instances using linear cartograms is highly advisable in practice; it is fast – in our implementation, it takes the time of about 50 annealing iterations on typical instances – and improves the process of finding a feasible initial solution in almost every aspect.

Incremental Greedy and Nonuniform sampling.   Finally we add the two augmentations from Section 6.3.1: additionally running the basic Incremental Greedy heuristic at every step, and nonuniform sampling of the vertex to move.

Still on the same instances, we first investigate on the additional executions of Incremental Greedy. We notice that Vertex-density annealing still finishes earlier on 644 instances (also significant, $z \approx 21.0$). On either procedure, the speed improvement obtained by adding Incremental Greedy is highly significant ($z > 36$ for both variants). Considering the movement cost of the found feasible drawings, we look at the average total movement for both variants: The average cost of all instances where Vertex-annealing finished first was 226.3 and the average cost of all instances where Grid-density annealing finished first was 188.9, a cost reduction by almost 50% in both cases. In fact, both cost improvements are also highly significant ($z > 50$ for both); the cost difference between both procedures is also significant ($z \approx 24.92$). Considering these significance levels, we decided to include Incremental Greedy executions into both of our Stage One procedures by default.

Since we now treat vertices differently, we investigate on the option of sampling vertices nonuniformly based on their local density measure. Looking at the average costs of the winning instances for both procedures, we get 196.2 for Grid-density and 229.9 for Vertex-density. Comparing these numbers to those with uniform sampling, one might think that nonuniform sampling is actually worse. Statistical tests however suggest that there is no significant difference between uniform and nonuniform vertex sampling; all

z-scores are below 0.34, implying that neither variant shows significant differences in quality or speed. We claim that the reason for this behavior the following:

Choosing the right vertex to try moving is a delicate task. On one hand, a vertex with low density is likely to have many moves that lead to valid drawings; on the other hand, none of these drawings will be significantly less dense (in either measure), as both density measures are somewhat symmetric: Vertices with high density values are likely to be close to other vertices of high density and the same is true for vertices with low density respectively. Thus, moving a low-density vertex has only limited impact on the overall density of the drawing. Successfully moving a high-density vertex would have a much higher impact on the overall density of the resulting drawing. This observation motivated trying nonuniform sampling. However, the higher the local density of a vertex, the harder it is to actually find a valid move for it. Hence, favoring the more dense vertices while sampling for possible moves is likely to result in more neighboring states being rejected by the topology test.

The final modification to the vertex sampling comes directly from how Incremental Greedy changes the network at every step. Recall that Incremental Greedy will try moving every nongrid vertex onto one of the four corners of the cell containing the vertex. Mutating a nongrid vertex in the local search neighborhood – moving it to a random corner of its cell – simply randomly tests one of the four options available to Incremental Greedy. Hence, after running Incremental Greedy on an instance, no nongrid vertex will have any valid move available to it (as it would have been performed by Incremental Greedy before). Therefore, every attempt of moving a nongrid vertex will immediately fail, rejecting the resulting state and effectively wasting the iteration. Instead we now only ever sample from the rounded vertices; if that results in creating a valid move for a nongrid vertex, this move will be performed by the execution of Greedy in the next round. Running the experiments on $(19, 40\%, 100\%)$ instances – complete Delaunay triangulations on 160 vertices – and testing for changes compared to sampling from all vertices, we failed to see a significant difference – all z-scores were below 1. Since we were under the impression that sampling only from the rounded vertices worked well on real-world instances, we look further into this. To do so, we created 1 000 instances of lower edge density, namely $(19, 40\%, 40\%)$, to resemble actual road networks. Performing the same experiments as before, we report the following findings: Vertex-density annealing got significantly faster ($z \approx 4.01$) while being just short of producing significantly cheaper feasible drawings ($z \approx 1.427$). On the other hand, the solutions produced by Grid-density annealing got significantly cheaper ($z \approx 3.30$), but we could not obtain the same speedup ($z \approx 0.389$). On these instances, we also observe that the speed difference between both Stage One procedures becomes insignificant ($z \approx 1.011$) while Grid-density annealing is still cheaper ($z \approx 17.85$). This also supports the claims we made when first comparing the two options in Section 6.4.2: indeed, Grid-density seems to thrive on easier and less dense instances.

Final Conclusion on Stage One. To sum up our findings on the different options for Stage One, we ran a final experiment on the 1 000 instances of $(19, 40\%, 100\%)$ with all options enabled: Cartogram preprocessing, Incremental Greedy at every iteration, and nonuniform vertex sampling ignoring nongrid vertices. Recall that the first sets of solutions had average costs of 347.208 for the Grid-density annealing and 407.916 for the Vertex-density annealing. All modifications described and evaluated above bring these numbers down to 183.082 and 221.572 respectively, saving almost 50% on the total cost of the first feasible solution and doing so in less time.

6.4.3 Evaluating Stage Two

Now that we demonstrated how to find a reasonable initial solution, we consider Stage Two. After Stage One departed from the input drawing to find a less dense but feasible drawing, Stage Two now anneals back towards the initial vertex positions. Annealing theory [KGV83] suggests that a system provided with the right cooling schedule and enough time will eventually end up in a low cost state. It is unclear how to determine that this state is reached, and a schedule that is guaranteed to achieve it with high probability is impractically slow; in practice, this requires experimental tuning.

We note that our moves from one drawing to the next are rather small, in terms of objective value: a single vertex is moved at most one grid cell, resulting is a maximum change in cost of $\sqrt{2}$ if moved diagonally. With the typical starting at temperature of $T_0 = 1.0$, this means we initially accept at least 24.3% of all cost-increasing moves. Similar to Section 6.4.2, we will evaluate various cooling schedules, step counts on large instances, and discuss when to switch from annealing to the hill climbing postprocessing as described in Section 6.3.3.

Finally, we note that unfortunately we cannot compare to optimal solutions for any interesting instances: As we have seen in Section 5.4, known exact methods are infeasible and only able to handle networks so small that the comparison to the output of our two-stage algorithm is not of interest – those instances are simply too small to give an intuition on how the output of our heuristic on real-world instances holds up to an optimal solution.

Cooling Schedules. One of the most important parameters for any simulated annealing process, besides number of iterations, is the cooling schedule. We use the typical exponential cooling schedule $T_{i+1} = c \cdot T_i$ which naturally gets slower over time. Therefore, we requiring a choice for initial temperature T_0 and cooling factor c. We evaluate this on 200 instances with parameters $(14, 40\%, 40\%)$ and 200 instances with parameters $(14, 40\%, 100\%)$. In order to eliminate any randomness from re-running Stage One, we use precomputed feasible solutions using Vertex-density annealing. We ran 20 000 steps in Stage Two (followed by hill climbing postprocessing) with $c_1 = 0.99$, $c_2 = 0.999$, and $c_3 = 0.9999$, doing five runs of each setting on each instance. Figure 6.5 shows a histogram of the resulting cost (rounded to the nearest integer). First consider Figures 6.5 (a) and (b), the instances with rather number of edges: before hill climbing (a), the choice of cooling factor does not seem to have high impact on the quality; afterward (b), the advantage of the slower cooling provided by $c = 0.9999$ becomes visible.

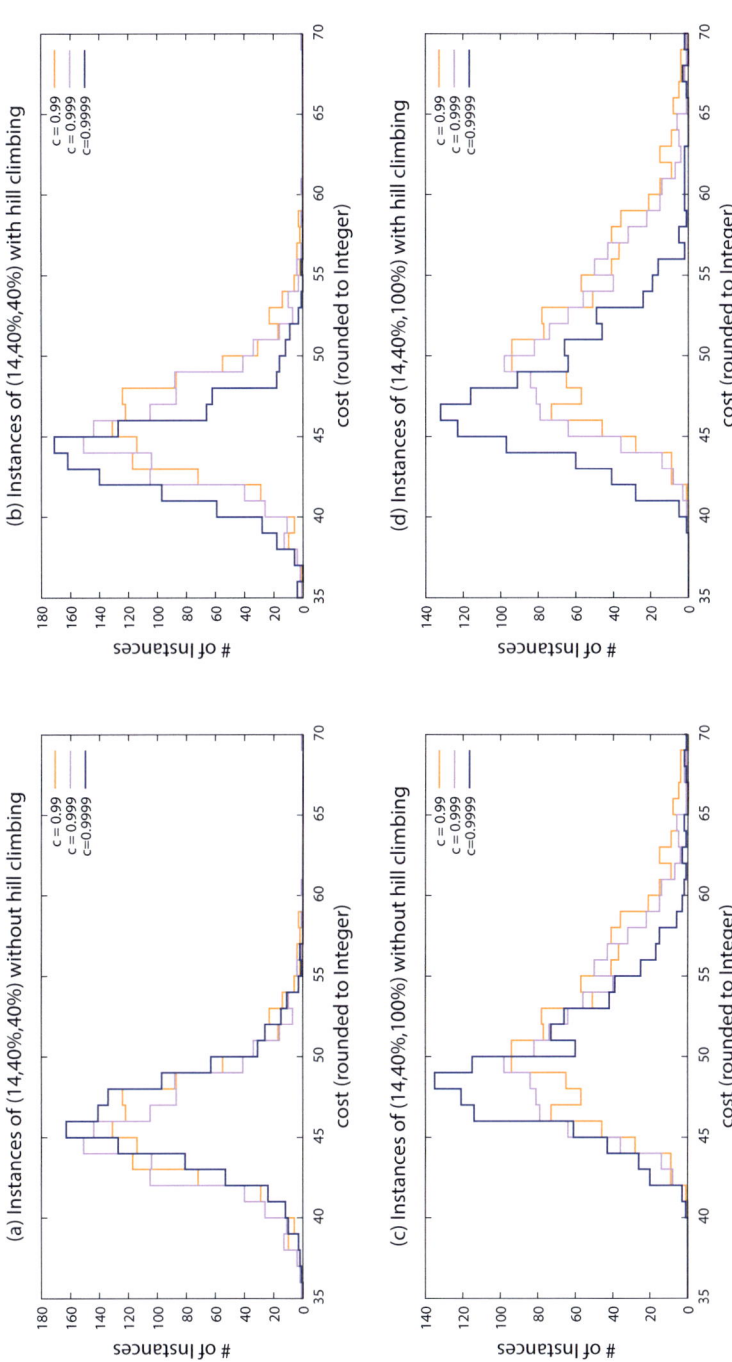

Figure 6.5: Effect of the cooling factor on final total cost. Each histogram combines five runs on 200 instances, results are packed into bins of integer size. The top row shows costs of (14, 40%, 40%) instances (a) before and (b) after hill climbing; the bottom row shows costs of (14, 40%, 100%) instances (c) before and (d) after hill climbing. (Single outliers beyond cost of 70 ignored.)

Looking at how the systems temperature developed over time, the reason is as follows: the schedule for $c_1 = 0.99$ reaches a temperature of effectively 0 after only 1500 iterations, compared to the 12000 iterations it takes for cooling with $c_2 = 0.999$ to reach the same temperature. This means that for both parameter values, a significant amount of time was spent rejecting any move that does not strictly improve the cost: they were basically hill climbing random vertices one step at a time. This behavior is also visible in the plots: There is hardly any difference between the orange and red lines in Subfigures 6.5 (a) and (b). This is not true for the slowest schedule ($c_3 = 0.9999$), as even after 20 000 iterations the temperature was still about 0.135 – while accepting is not very likely in the end, this schedule never rejected score-decreasing moves immediately, leaving room for improvements to be made by hill climbing. The same phenomenon can be observed looking at the second set of instances (Figure 6.5 (c) and (d)). Again, hill climbing basically did not find any improvements for c_1 or c_2, whereas c_3 managed to avoid running into local optima long enough for hill climbing to make a difference. The cost improvement of hill climbing on both test sets are highly significant ($z > 50$ each).

With these observations in mind, we recommend always using the hill climbing as postprocessing; it is fast and guaranteed to not worsen the final solution as hill climbing only performs a set of moves that would have been accepted by annealing even at temperature 0.

Step Count & Hill Climbing. To further evaluate the effect of hill climbing postprocessing, we generated feasible solutions for all of the $(19, 40\%, 100\%)$ instances from Section 6.4.2 using Vertex-density annealing with cartogram preprocessing.

On these rather large and dense instances, we run Stage Two annealing for m steps ($m \in \{0, 2\,500, 5\,000, 10\,000, \ldots, 40\,000\}$), followed by hill climbing, reporting score with and without postprocessing. To demonstrate the impact of step count – and with respect to the experiments above –, we choose the slowest of the cooling schedules above, namely $c = 0.9999$. Indicative results of these experiments are shown in Figure 6.6.

We deal with instances of maximum edge density that result in rather expensive Stage One solutions (see Section 6.4.2). This leaves quite some space for a lot of improvements – we can expect "good" final solutions to have cost below 100, or less than 1 per vertex on average. We now discuss the four subfigures of Figure 6.6 individually.

As noted, annealing with temperature 0 yields very similar results to hill climbing. After 5 000 steps, the system is still at temperature 0.607, whereas after 40 000 steps, it reaches a temperature of 0.0183. Also recall that high temperatures imply high acceptance probability for Stage Two (between 24.5% and 16.4% for the first 2 500 steps). A downside of combining a slow cooling schedule with greedy local optimization is visible in Figure 6.6 (a). For the orange curve, we spent 2 500 iterations randomly moving away from the first feasible solution, which hill climbing sometimes could not repair.

Figure 6.6 (b) compares $m = 15\,000$ steps with hill climbing to $m = 25\,000$ steps without postprocessing. For reference, we also include the data for $m = 2\,500$ from Figure (a). Notice that performing 22 500 extra steps in Stage Two leads to lower costs in the final drawing, even without hill climbing. However, carefully picking the point

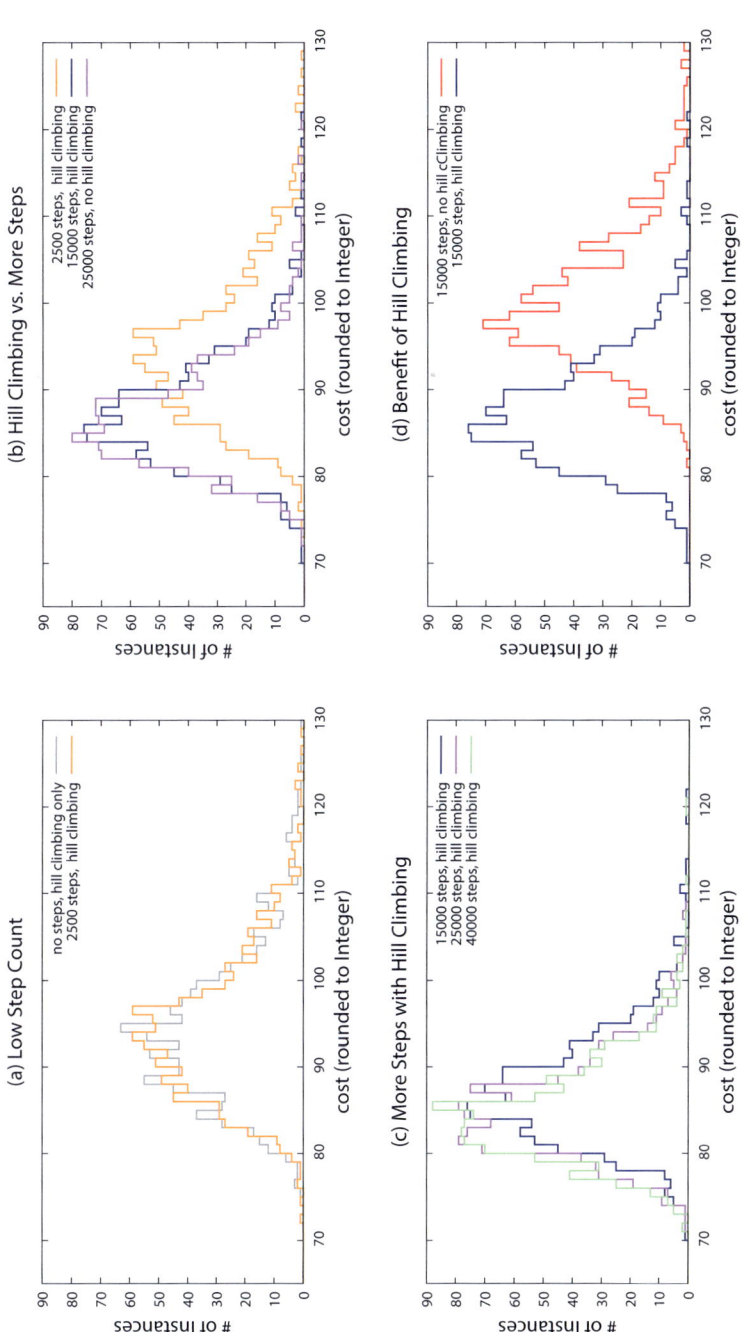

Figure 6.6: Various values of step count m for Stage Two and the effect on cost of $(19, 40\%, 100\%)$ instances with and without hill climbing: (a) $m = 0$ vs. $m = 2\,500$, both with hill climbing. (b) The trade-off between hill climbing and running Stage Two for more steps. (c) Multiple values for $m \in \{15\,000, 25\,000, 40\,000\}$, with hill climbing. (d) $m = 15\,000$, with and without hill climbing. Some outliers have been cropped; for easier comparison, line colors correspond to datasets and carry over into other plots.

at which to switch from annealing to hill climbing can lead to similar results in shorter time, as shown by the blue line.

Figure 6.6 (c) shows how the total movement cost changes with increasing step count. There is a noticeable difference between high and low step counts. In addition, even after 40 000 annealing iterations, hill climbing managed to improve 874 of the 1 000 instances.

Figure 6.6 (d) illustrates the impact hill climbing can have on the total cost of a solution. The improvement made by hill climbing is obvious and easy to explain – Stage Two simply was not done yet. Nevertheless, the final result after postprocessing is comparable to those of higher iteration counts.

6.4.4 Real-World Data

In this section we compare instances based on real-world road networks to artificial networks. The results presented here were obtained by experiments we ran on a 2.6 GHz processor with sufficient RAM. See Table 6.2 for instance parameters and sizes as well as experimental results; all data was gathered over 100 runs on each instance, reporting average scores and times, together with standard deviations over these measurements. More data on other instances is publicly available online[12].

The instances were preprocessed using linear cartograms and Vertex-density annealing with all options enabled for Stage One. Examine the instances and the computed grid drawings of Würzburg–Train Station, found in Figure 6.7 (a). The solutions of Würzburg–Train Station we computed had an average cost of 110.9 (0.860 on average per vertex, standard deviation of 8.66); notice that for these runs, we picked the vertex-grid point ratio γ = 16.45%. The average cost over 100 runs on artificial instance named "0.4_0.4_42" (resembling this real network by picking ε accordingly, but with γ = 40%) is 77.2 (0.483 on average per vertex, standard deviation of 1.92). While Würzburg–Train Station is less dense and thus could be expected to be easier, neither cost nor runtime reflect this; Train Station takes 63% longer on average (43.1 s vs. 26.4 s) and is 43% more expensive. To investigate this, consider that the left part of Fig 6.7 (a) (marked by the blue arrow) is significantly more dense than the rest; moving the tilted bus parking lanes to the grid results in parts of them being pushed outwards. The vertex indicated by the blue arrow (together with the middle of its incident edges) forces significant distortion in this drawing. While there seem to be quite some empty grid points nearby, topology prohibits any local improvement for the vertex marked by the orange arrow and on closer reflection, this seems to be a decent drawing of this instance on a critically small grid. To support this observation, we extracted the dense part: Würzburg–Bus Lanes alone at γ = 40% is comparable to artificial instances of about twice its size on runtime and rounding cost.

This problem is even more obvious for Chicago–Cloud Gate, shown in Figure 6.7 (c). Note that the path to the bottom-right of the instance has way too many vertices compared to the grid size. Hence we get a rather skewed grid representation even though the

---

[12] http://github.com/tcvdijk/armstrong

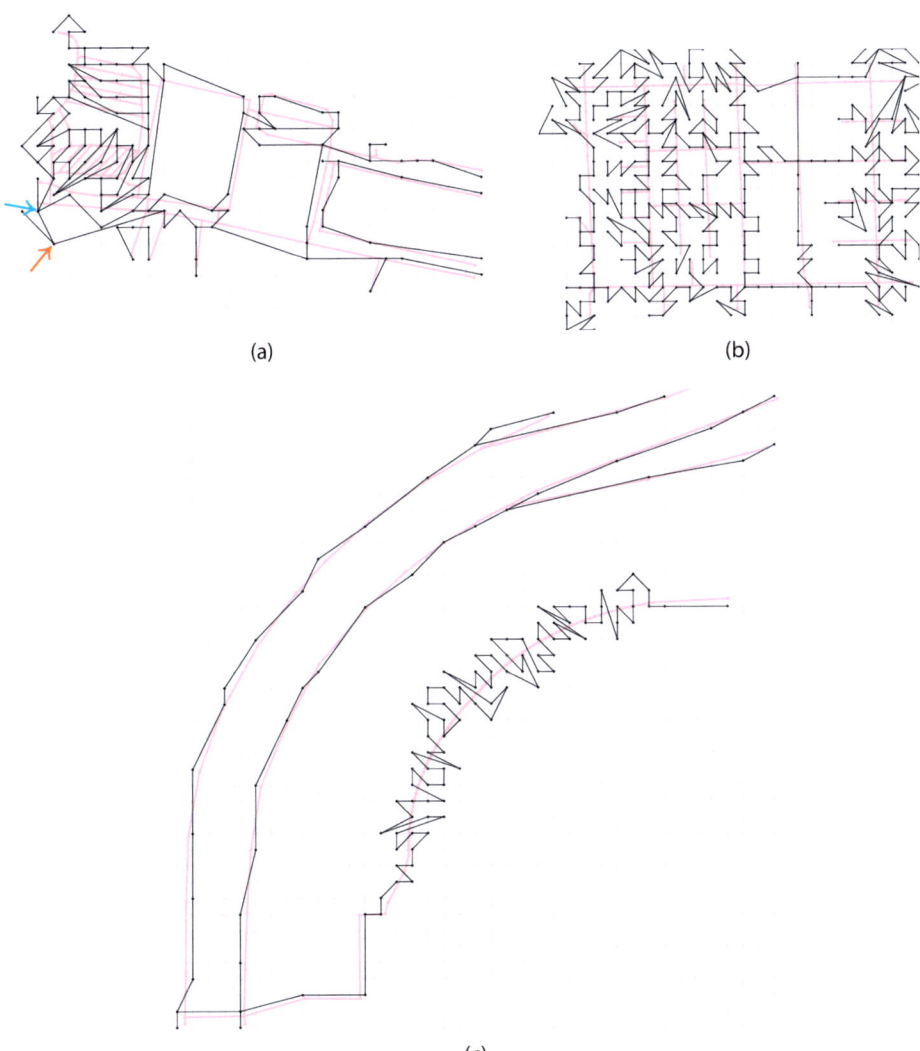

(a)

(b)

(c)

Figure 6.7: Grid representations (in black) of real-world instances (in red). (a) Würzburg–Train Station on a grid of size 28 × 28; the vertex indicated by the orange arrow looks highly suboptimal. (b) Chicago–Downtown on a grid of size 25 × 25; (c) a crop of Chicago–Cloud Gate on a grid of size 240 × 240.

Table 6.2: Selected real and artificial instances with experimental results; all numbers are the average over 100 runs with the same parameters. "Runtime" is given in seconds, single-threaded; "m" is the number of iterations in Stage Two; "S1 cost" and "S2 cost" are total rounding cost after Stages One and Two respectively; σ_t, σ_{S1} and σ_{S2} are the corresponding standard deviations on the measurements. Würzburg–Train Station is shown in Figure 6.7 (a), Bus Station is the left half of Train Station; Chicago–Downtown and –Cloud Gate can be found in Figure 6.7 (b) and (c). Würzburg–Ring and parts of United Kingdom–Borders are shown in Figure 6.1 (a) and (b) respectively (Würzburg–Ring is shown on a finer grid of size 28 × 28).

Name	v	ε	size	γ	m	time	(σ_t)	S1 cost	(σ_{S1})	S2 cost	(σ_{S2})
Wü –Train Station	129	40.3%	28 × 28	16.5%	20 000	43.2 s	(14.90 s)	486.2	(119.9)	110.4	(7.20)
–Bus Station	89	41.7%	15 × 15	39.6%	20 000	23.9 s	(4.73 s)	242.0	(59.7)	72.6	(6.23)
–Ring	138	38.8%	20 × 20	34.5%	20 000	60.7 s	(9.54 s)	392.5	(67.2)	115.4	(4.94)
Chi –Downtown	358	35.2%	25 × 25	57.3%	20 000	83.3 s	(18.30 s)	1462.7	(106.8)	390.4	(16.0)
–Cloud Gate	578	35.1%	240 × 240	1.0%	35 000	492.0 s	(119.40 s)	2 090.8	(313.1)	356.6	(8.69)
UK –Borders	3 110	34.8%	250 × 250	5.0%	50 000	590.0 s	(282.30 s)	5 817.3	(938.5)	1 344.2	(11.1)
Art –0.4_0.4_42	160	40%	20 × 20	40.0%	20 000	26.4 s	(0.88 s)	135.7	(0)	77.2	(2.15)
–0.4_0.4_84	160	40%	20 × 20	40.0%	20 000	25.6 s	(1.50 s)	205.9	(50.4)	77.9	(3.18)
–0.4_1.0_42	160	100%	20 × 20	40.0%	20 000	63.7 s	(11.50 s)	454.6	(181.9)	94.7	(10.1)

other roads around the problematic path are generally represented quite well (and would have tolerated an even coarser grid). The average vertex cost after Stage One (3.62 per vertex) indicates how long it took to find any feasible drawing. This suggests that future work could attempt to integrate topologically-safe simplification into the grid representation workflow.

Consider the artificial instances 0.4_0.4_42 and 0.4_0.4_84, again from Table 6.2: different networks with the same parameters. The former is immediately solved by Incremental Greedy during the first iteration of Stage One, so the result is deterministic. For this instance, this was only the case when cartogram preprocessing was enabled, demonstrating its benefit. Though the average runtime on the two instances is comparable, the final drawings of the latter have higher variance. This suggests that a deterministic Stage One may be preferable, as the results obtained from Stage Two will become more stable.

All of this indicates that the performance of our algorithm is sensitive to the structure of the network, and that the particular road network representations we found in OpenStreetMap and the City of Chicago Data Open Data Portal are challenging instances. Still, our method is able to find reasonable representations of realistic networks, which was not feasible with previous methods.

6.5 Conclusion

In this chapter, we settled one of the open questions from Chapter 5: We provided an efficient heuristic to the TOPOLOGICALLY-SAFE GRID REPRESENTATION problem. It relaxes on the minimality of the total vertex displacement in the output: finding non-optimal, yet reasonable solutions, while maintaining topological equivalence of input and output drawing. This heuristic was designed to transform geographic networks into drawings requiring lower coordinate precision. To do so, we proposed a two-stage approach based on the simulated annealing metaheuristic. The first stage anneals for feasibility by reducing the overall vertex-density of the drawing; the second stage anneals the feasible drawing to reduce the total rounding cost of the output.

We demonstrated the necessity of our two-stage approach by experimental evaluation. For Stage One, we proposed two different objective functions for annealing, and by analyzing their performance on randomly generated artificial road networks, we were able to provide significance tests to show the strong and weak points of both. For Stage Two, we provided experiments demonstrating the impact of parameter selection to the final result. We proposed pre- and postprocessing steps and showed via significance testing that both procedures improve the result of their respective stage significantly.

We concluded the experimental evaluation of our algorithm by testing it on handpicked real world instances of various sizes.

Future research on the topic of TOPOLOGICALLY-SAFE GRID REPRESENTATION could revolve around finding a deterministic heuristic or an approximation algorithm with provable guarantee. It could also be beneficial to the algorithms output to look into moving the input drawing onto a non-uniform grid. One could, for example, imagine distinguishing between the downtown- and highway-parts of a road network. With the application to road networks in mind, one could also consider using other techniques (such as line simplification) to better preserve the visuals of the input in the output drawing while reducing unnecessary detail.

Acknowledgments. We thank Thomas C. van Dijk for helping with creating sophisticated implementations of the procedures described in this chapter in C++, and for making datasets and the implementation publicly available at
`https://github.com/tcvdijk/armstrong`.

Chapter 7

Cauchy's Theorem for Orthogonal Polyhedra

A classic theorem by Cauchy states that, for convex polyhedra, when the embedded graph of the surface and the angles within each face are given, then the dihedral angles are determined – that the object is rigid. In this chapter, we translate Cauchy's rigidity theorem to orthogonal polyhedra of arbitrary genus. We do so by using the linear-time BundleOrientation algorithm by Biedl and Genç [BG09] as a subroutine. They originally created it to determine the unique set of dihedral angles of orthogonal polyhedral surfaces of genus 0 with connected graph (if this set exists).

They left open whether a similar translation exists for orthogonal polyhedral surfaces of higher genus. In this chapter, we resolve this in the affirmative. To obtain this result, we apply the original BundleOrientation algorithm repeatedly and exhaustively, and call this procedure IteratedBundleColoring. We show that it is capable of finding a set of dihedral angles for orthogonal polyhedral surfaces of arbitrary genus. We do so by arguing how it re-discovers the dihedral angles matching those of a polyhedron realizing the input graph.

Concepts. We begin this chapter with formally defining the structures we consider.

A *polyhedron* is a solid object in three dimensions. Looking at a polyhedron, we see straight edges and sharp point-shaped corners on its exterior. The edges form a set of induced cycles, creating two-dimensional polygons on the surface. These polygons define the *faces* of the polyhedron. Joining all polygons, we obtain the polyhedral surface bounding it – the *genus* of a polyhedron is the genus of the bounding surface.

A polyhedron induces a combinatorially (and geometrically) embedded graph via its boundary – the faces, edges, and vertices of the graph correspond to those defined by the corners and creases of the polyhedron. Moreover, the relative placement and orientation of the connected components in the graph are specified by the polyhedron. Similarly, a *polyhedral surface* in three dimensions is a closed connected orientable mesh of polygonal faces – the difference is that adjacent faces can be parallel. A polyhedral surface also induces an embedded graph, which we call a *net*. By convention, nets are connected. Note that, as remarked by Ziegler [Zie08], the classification of polyhedra and polyhedral surfaces up to homeomorphism is well-known: For each integer $g \geq 0$, there

---

This is joint work with Steven Chaplick and Thomas C. van Dijk.

is exactly one topological type, *the surface of genus g*, obtainable by attaching g handles to the 2-sphere $\mathbb{S}^2$.

Given a single polygonal face of a graph, the *facial angles* are the angles at which two consecutive edges meet; for two adjacent flat polygonal faces in three-dimensional space, the *dihedral angle* is the angle measured along the edges they share.

A classic topic in geometry is polyhedra (and polyhedral surfaces) with various geometric restrictions, the most well-studied case being the *convex* polyhedra where the vertices occur in convex position. For example, convex polyedra were a common topic of Euclid, Cauchy, and Steinitz. Recently several classic results on convex polyhedra have been considered for *orthogonal* polyhedral surfaces, for which every edge is parallel to one coordinate-axis; here, all facial and dihedral angles are multiples of $90°$, and (without loss of generality) each face is perpendicular to one coordinate-axis.

Fixing the edge lengths of the graph to be at integer precision, any orthogonal polyhedron realizing the graph with those edge lengths can naturally be "snapped" to the three-dimensional integer grid. This can be done by picking any face of the realization, making its edges parallel to two of the coordinate axes – by rotating the polyhedron – and then shifting the polyhedron to make one of the face's corners coincide with the coordinate systems origin. Given an edge of the polyhedron with one endpoint on the integer grid, orthogonality and integer edge lengths ensure that the other endpoint has to be on the integer grid as well.

Next, we outline some results from the literature.

7.1   Related Work and Contribution

Classical Convex Results vs. Recent Orthogonal Results.   *Cauchy's rigidity theorem* states that, when constructing a convex polyhedron from a 3-connected planar graph, the facial angles determine the dihedral angles uniquely; proofs can be found in several textbooks, for example, see *Proofs from THE BOOK* by Aigner and Ziegler [AZ04].

This theorem breaks down for non-convex polyhedra. Consider an object composed of two six-sided polyhedra – one large and one small – with the smaller one attached to one side of the larger one (the faces they are connected at are coplanar and the smaller face is contained in the interior of the larger face). Notice that, even if given the net, the facial angles, and edge lengths of this combined polyhedron, it is still not uniquely described – it can be realized as a polyhedron in the two distinct ways as shown in Figure 7.1: adding the smaller polyhedron to the larger (a), creating a "bulge", or subtracting the smaller polyhedron (b), creating a dent. Both variants are valid nonconvex realizations, creating different dihedral angles. Notice that both realizations create the same graph via their corners and creases, see Figure 7.1 (c).

Moreover, there are even *flexible* polyhedra where the dihedral angles can vary continuously while maintaining the edge lengths or facial angles; one such object was created by Connelly [Con79]. The Bellows conjecture claims that flexible polyhedra also have fixed volumes. The conjecture was shown to be true for general orientable 2-dimensional poly-

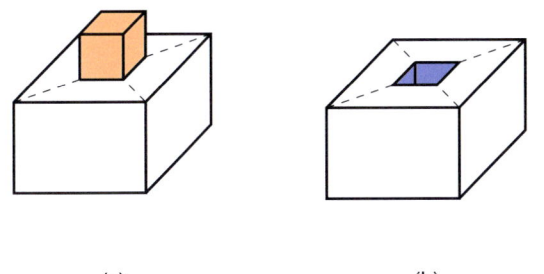
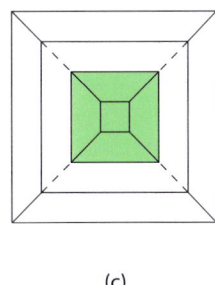

(a)   (b)   (c)

Figure 7.1: Example of a non-convex polyhedron that is not rigid: (a and b) Two possible non-convex realization for the graph from (c) with the green structure "pushed out" (in orange) (a) or "dented in" (in blue) (b). (c) The net of both polyhedra is disconnected – the green piece is created by either the orange or blue piece (dashed lines indicate the relative position on the surface).

hedral surfaces by Connelly, Sabitov, and Walz [CSW97]. Of course, orthogonal polyhedra are inherently inflexible.

An analogous formulation of Cauchy's rigidity theorem for orthogonal polyhedral surfaces of genus 0 is shown by Biedl and Genç [BG09] where the graph is connected[1]. Note that, like convexity, the connectedness is necessary for the uniqueness of the dihedral angles. This is illustrated in Figure 7.1 (b) and (c): When the two joining faces (indicated by the dashed connections) touch in one edge – connecting the graph –, subtracting the smaller polyhedron would create a degenerate object that has a "boundary-only" section without interior at the other side of the joined edge. In an extended version [BG08] of their work [BG09], Biedl and Genç show that testing whether a given embedded planar graph can be realized as an orthogonal polyhedral surface where every face is a unit square is $\mathcal{NP}$-complete. The gadgets utilize disconnectedness of the graph.

Steinitz' theorem states that the 3-connected planar graphs are precisely the graphs obtainable from the surfaces of convex polyhedra. A graph-theoretic characterization of orthogonal polyhedra of genus 0 where three mutually-perpendicular edges meet at each vertex has also been proven by Eppstein and Mumford [EM14].

Stoker's theorem states that, when constructing a convex polyhedron from a 3-connected planar graph, the dihedral angles and edge lengths determine the facial angles uniquely. However, Bield and Genç [BG11] showed that for embedded graphs with given edge lengths and dihedral angles, it is $\mathcal{NP}$-hard to decide if there is an orthogonal polyhedral surface of genus 0 realizing this input.

---

[1] Recall that the graph of the polyhedron is defined by the corners and creases, and as all dihedral angles are non-zero, there are no coplanar faces incident to the same edge.

A related topic is whether any given embedded graph having facial angles restricted to multiples of 90° forces all of its realizations as polyhedral surfaces to be orthogonal. For example, Biedl, Lubiw, and Sun [BLS05] asked whether every polyhedron in which every face is a rectangle is an orthogonal polyhedral surface. Their question was answered in the negative by counterexamples of genus 7 (Donoso and O'Rourke [DO02]) and genus 6 (Biedl et al. [BCD$^+$02]. On the other hand, the question has a positive answer for genus at most 2 [BCD$^+$02, DO02]. The cases of genus 3, 4, and 5 remain open.

Our main result – stated in Theorem 7.1 below – builds upon the work of Biedl and Genç [BG09] from orthogonal polyhedral surfaces of genus 0 to all orthogonal polyhedral surfaces. They already give the following two-step strategy that we extend upon in this chapter.

In the first step, they show that for each face f, the axis perpendicular to f is uniquely determined by a given embedded *genus-0 (planar)* graph with specified facial angles (up to relabeling of the axes). Note that this first step already determines all of the "flat" dihedral angles, i.e., the parallel adjacent faces, and as such provides the unique graph of every realization (obtained by merging parallel adjacent faces).

The second step consists of two small substeps. They first observe that connectedness of the graph and knowing the axis perpendicular to each face implies that there can be only two possible realizations of the dihedral angles, basically regarding which side of the faces is the "outside" and which side is the "inside" of the surface. With this observation in mind, they further note that having fixed edge lengths fully determines all of the vertex positions in both sets of dihedral angles (up to translation).[2] Thus, from this set of vertex coordinates they identify the correct set of dihedral angles by looking at the incident dihedral angles of a face f perpendicular to the x-axis with the maximum x-coordinate (f's dihedral angles must all be 270°). Finally, they note that quadratic time suffices to check that the constructed object avoids self-intersections using an existing algorithm [BLS05]. This resolves the genus-0 case.

Remark 7.1. Interestingly, as stated by Biedl and Genç [BG09], the second step does not require a genus-0 input, and can be applied (regardless of genus) as long as one is given the axis perpendicular to each face. Thus, to generalize their result from genus 0 to arbitrary genus, it suffices to generalize the first step of their approach.

They give the original BundleOrientation algorithm – a conservative propagation algorithm working on the edges of the graph, which we describe in Section 7.2 – and show that (in the case of a realizable input) its output uniquely determines the axis perpendicular to each face with respect to any *starting face*. They demonstrate that their approach can fail to determine the axis for every face already for realizable inputs of genus 1 (see Figure 7.2 (a) and further details in Section 7.2). In particular, they observe that for some starting faces, their algorithm does not determine the axis of every face. Additionally, we present an example of genus 2 in Figure 7.2 (b) where no single face is a sufficient place to start to resolve the axis of every face via their algorithm. We discuss this in Section 7.2.

---

[2] This can result in one vertex needing two distinct positions, in which case the instance is not realizable.

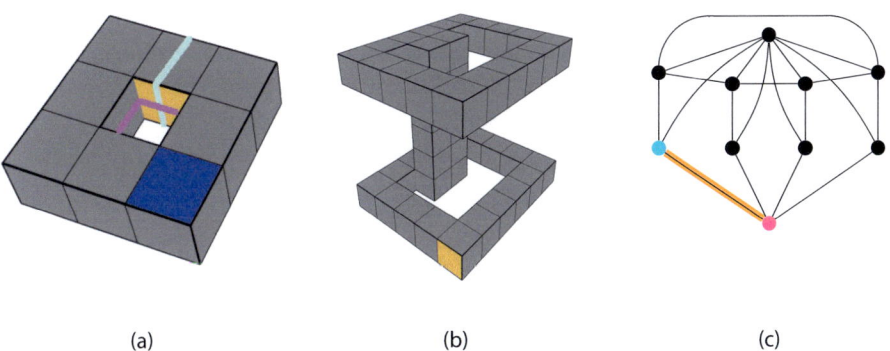

(a)   (b)   (c)

Figure 7.2: (a) A polyhedron of genus 1 with a bad starting face in yellow (one of four). The face's bundles (in purple and turquoise) cross only once, stopping propagation. (b) A polyhedron of genus 2 composed of two tori connected by a bridge going from the inside of one to the inside of the other. Starting at the yellow face orients the bottom torus but not the top one. (c) The underlying bundle graph $\mathfrak{B}_G$ of the polyhedron from (a), with matching colors identifying the same objects.

Contribution. The main result of this chapter is generalizing the main theorem of Biedl and Genç to orthogonal polyhedra of arbitrary genus. To do so, we prove the following extended theorem:

Theorem 7.1. *Given an embedded graph G with facial angles F and edge lengths L*

1. *we can report in cubic time either that no orthogonal polyhedral surface realizes the given graph and facial angles or determine the unique coordinate axis perpendicular to each face; and*

2. *if 1 does not reject the instance, then in additional linear time we report:*

   - *that this graph and facial angles can only belong to an orthogonal polyhedral surface for which the polyhedron bounded by it has a disconnected graph; OR*
   - *that no orthogonal polyhedral surface realizes (G, F, L); OR*
   - *the unique set of dihedral angles of any orthogonal polyhedral surface that has this graph as its net, facial angles F, and edge lengths L.*

We overcome the difficulty of finding the right starting faces by describing how to use their BundleOrientation algorithm in an exhaustive fashion – obtaining the algorithm we call IteratedBundleColoring. It progressively learns the orientations of faces, see Section 7.2. The correctness of our approach is presented in Section 7.3. There we consider an input graph and facial angles together with a hypothetical polyhedron that realizes it. We traverse the surface of this realization "layer-by-layer", maintaining the invariant that IteratedBundleColoring would orient the faces of the next layer (by coloring the bundles that their edges belong to), if it could orient the faces of the earlier layers. This invariant is stated in Lemma 7.4, which we will split into sub-cases for all possible different two-layer patters, providing and individual lemma for each. By arguing that the output of IterateDBundleColoring is *stable* – that is, it is unique up to renaming the equivalence classes –, we can use this output directly as input for the original second step by Biedl and Genç, implying correctness of Theorem 7.1.

7.2   Orienting Faces by Coloring Edges

In this section we provide the IteratedBundleColoring algorithm to accomplish the following task. The input of the algorithm is a connected and embedded graph G (i.e., including the cyclic order of the edges around each vertex, and the corresponding faces) as well as the set F of facial angles, each of which is a multiple of $90°$. If there is an orthogonal polyhedral surface realizing (G, F), then the algorithm will determine an *orientation* of each face, i.e., for each face f, an axis perpendicular to f or, equivalently, the *axis-plane* parallel to f will be specified. Moreover the orientations are obtained in a *stable* way: Up to renaming of the axes, every orthogonal polyhedral surface realizing (G, F) has these orientations.

The algorithm builds upon the reasoning used by Biedl and Genç [BG09] for the case of genus 0. Rather than directly working with the faces of G, it works on the *bundle graph* $\mathfrak{B}_G$. For any orthogonal face f, the edges bounding f can be partitioned into two *(edge) bundles*. If a pair of edges is parallel (with respect to parity and the facial angles at f's corners), they are in the same bundle; otherwise, they are in different bundles. This implies that the edges of each face are partitioned into two bundles where edges in distinct bundles are perpendicular. If edges of two distinct bundles appear on the same face, we say that those bundles *cross* there. By definition, each edge e bounds exactly two faces and thus is in two (possibly) different bundles. As parallelism is transitive, all edges of both such bundles are parallel to e, allowing the bundles to *merge* into one. Taking the closure of these merges leads to a partition of the entire edge set of G into a collection $\mathcal{B}$ of bundles where edges of distinct crossing bundles are perpendicular in any realization as an orthogonal polyhedral surface. This gives us the *bundle graph* $\mathfrak{B}_G = (\mathcal{B}, \mathcal{E})$ with vertex set $\mathcal{B}$ and an edge $e \in \mathcal{E}$ between two vertices, if there is at least one face containing edges of both bundles.

To discover the realization as an orthogonal polyhedral surface, we aim to partition the edges into three color classes depending on the coordinate-axis to which they are parallel. BundleColoring picks an edge of $\mathfrak{B}_G$ (a face of G) and colors the two bundles connected by it differently. It then repeatedly looks for a triangle in $\mathfrak{B}_G$ with two colored vertices, coloring the third vertex in the remaining color, until all of $\mathfrak{B}_G$ is colored or no such triangle is found anymore. As a shorthand, we use BundleColoring($\mathfrak{B}_G, e$) to say we run BundleColoring on the bundle graph $\mathfrak{B}_G$ with starting edge $e \in \mathcal{E}$. Then, any 3-coloring of $\mathfrak{B}_G$ provides an orientation of the faces of G.

Algorithm 7.1: ITERATEDBUNDLECOLORING(Graph G, facial angles F)

$\mathfrak{B}_G = (\mathcal{B}, \mathcal{E}) \leftarrow$ bundle graph of G with facial angles F
while $|\mathcal{B}| > 3$ **do**
    $\mathfrak{B}_G^{old} \leftarrow$ copy of $\mathfrak{B}_G$ `/* Copy to track progress */`
    foreach *edge* $e \in \mathcal{E}$ **do**
        merge bundles in $\mathfrak{B}_G$ using BUNDLECOLORING($\mathfrak{B}_G, e$)
    `/* If nothing changed and there are still uncolored`
        `bundles, subsequent runs will also fail.`   `*/`
    if $\mathfrak{B}_G^{old} = \mathfrak{B}_G$ **then return** Infeasible
return $\mathcal{B}$ `/*` $\mathcal{B}$ `is now a partition of the edges of` G. `*/`

Remark 7.2. If the input graph G consists purely of rectangular faces, then the corresponding bundle graph has a natural intuitive structure. Each bundle corresponds to a cycle of faces obtained by tracing the outline of the surface along a path parallel to one of the axes – see Figure 7.2 (b) for the surface and (d) for the corresponding bundle graph (using the same colors).

They showed that for realizable instances (G, F) where G has genus 0, this simple conservative coloring procedure always completely colors $\mathfrak{B}_G$, and hence the orientation of the faces in any realization of (G, F) is unique up to naming of the axes. Their algorithm is easily implementable in time linear in the size of the bundle graph, which is linear in the size of G. Biedl and Genç observed that BUNDLECOLORING can fail to color all the bundles when the graph G has genus 1. For example, if, in Figure 7.2 (a), we start the bundle coloring procedure on the edge corresponding to the (yellow) "inner" face, then it will not color any bundles beyond the original two. However, if a "corner" face (e.g. the blue face) is used instead, then all bundles will indeed be colored. As their algorithm starts with an arbitrary edge, it cannot generally cope with such examples. Therefore, Biedl and Genç left the status of nonzero-genus inputs as an open problem. In fact, already for genus 2 we observe that there are realizable instances where some bundles will remain uncolored after one execution of BUNDLECOLORING regardless of the starting edge. For example, in Figure 7.2 (b), if we start from a face of one torus (e.g., the yellow face), when the color propagation reaches the other torus, the topology steers it so that it is as though we started on an inner face of the other torus. Thus, as in the genus-1 example, some bundles remain uncolored. This implies that a more involved approach is needed for genus larger than one. Here, we can observe that the faces of this "bridge" connecting the torii will be oriented when starting in a "good" face for either torus. intuitively, by combining these two partial orientations, we can indeed consistently orient all of the faces.

For the case of arbitrary genus inputs, we designed the ITERATEDBUNDLECOLORING algorithm, listed in Algorithm 7.1. It first constructs the bundle graph from (G, F) and then repeatedly runs rounds of BUNDLECOLORING executions on the edges of the bundle graph as a subroutine. While doing so, it intermediately adjusts the bundle graph depending on the information learned from the previous runs. Namely, if a single run of BUNDLECOLORING($\mathfrak{B}_G, e$) on some edge $e \in \mathcal{E}$ does not color all of the bundles, but does manage to color some bundles, we can derive the following: All bundles that have received the same color must do so in any valid 3-coloring. Thus, for each $i \in \{1, 2, 3\}$, we merge the bundles with color i into a single bundle where the neighborhood of this bundle is simply the union of the neighborhoods of its members. The ITERATEDBUNDLE-COLORING algorithm simply repeats such rounds until no bundles are merged in a round (since this implies that no further iterations would merge bundles). For the total runtime of ITERATEDBUNDLECOLORING, we have the following lemma:

Lemma 7.2. *On an m-edge graph G, ITERATEDBUNDLECOLORING runs in $O(m^3)$ time.*

Proof. Consider the total number of merge operations using BUNDLECOLORING. They will occur at most as many times as we have bundles to begin with, i.e., less than the number of edges in G. On a graph with n vertices and m edges an execution of BUNDLE-COLORING takes $O(n+m)$ time. In each round, we run BUNDLECOLORING for every face (of which there are $O(m)$). Thus, after at most linearly many rounds, we will indeed have a round in which no bundles merge, resulting in a total runtime cubic in the number of edges. □

In Section 7.3, we will show that for realizable instances (G, F) of arbitrary genus, this simple iterated conservative coloring procedure results in a triangle (i.e., it completely colors and merges the bundles), and as such the orientation of each the face in any realization of (G, F) is unique up to naming of the axes. This will establish Theorem 7.1.

Remark 7.3. For the correctness of our approach, it suffices to consider input instances where each face is a unit square. In particular, when an instance (G, F) can be realized, the resulting orthogonal polyhedral surface can be tessellated so that every face is a unit square, providing a corresponding tessellation G' of G. Moreover, the bundle graph of G' only contains less information (more separate bundles) regarding which edges of G must be parallel and which must be perpendicular. So, by arguing that the edges of any such tessellation will be completely colored by our approach, we indeed establish that the edges of G will also be completely colored.

7.3 Arbitrary Genus: The Proof of Theorem 7.1

Following Remark 7.1 given by Biedl and Genç [BG09] on their two-step proof structure, to prove Theorem 7.1, it suffices to prove the following theorem regarding step one.

Theorem 7.3. *Given a connected embedded graph G (of arbitrary genus) with facial angles F that are all multiples of $90°$, ITERATEDBUNDLECOLORING will*

- *report that no orthogonal polyhedral surface can realize this graph and facial angles, OR*

- report all edges of the graph for which the dihedral angles must be $180°$ in any orthogonal polyhedral surface that realizes this graph and facial angles.

Recall that, by Remark 7.3, it suffices to prove this when every face of G is a unit square. With this in mind, we set up some terminology. Let G be a graph in which all faces are bounded by cycles of length four, and let F be a set of facial angles that are all $90°$. Together G and F imply that the surface of any orthogonal polyhedron P realizing (G, F) has to be tessellated using rectangular faces. Imposing the additional restriction for every edge to have unit-length, and that the faces are supposed meet at dihedral angles that are multiples of $90°$ implies that P itself – if it exists – needs to be composed of solid unit cubes joined at their sides (not edges or corners).

7.3.1 Proof Outline

In the following we assume that (G, F) is a realizable instance and that P is the polyhedron realizing it. We will traverse the polyhedron in a top-to-bottom fashion, considering locally confined pieces of G. Each piece will correspond to a subset of the faces on P and we will argue that IterateBundleColoring would discover the orientations of those faces assuming the higher components of P are fully oriented. This process will yield that the orientation found on P is unique (up to rotation), because BundleColoring will only ever derive the coloring of a bundle when it shares faces with bundles of the two other colors.

Traversing Object and Graph. We assume that P is aligned to some three-dimensional coordinate system: At least one of the faces of P is coplanar to the xy-plane π_0 at the origin[3] – imagine P standing on that face, growing upwards. Let t be the maximum z-coordinate of all points of P. By definition, the "top-most" faces of P (having an interior point with z-coordinate t) are coplanar to π_t. Similarly, all "bottom-most" faces (with an interior point with z-coordinate 0) are coplanar to π_0. Furthermore, any plane π_i with $i \in (0, t)$ crosses through the interior of P. This allows us to define the i-th cross section C_i of P, the set of cubes intersecting π_i. Obviously, C_x is empty for $x \notin [0, t]$. The term "cross section" is a misnomer here, as those are usually two-dimensional outlines created by intersecting a three-dimensional object with some plane, and not also three-dimensional objects. In fact, we will later also consider the two-dimensional outlines of those objects (projecting the cubes onto the xy-plane they are intersected by).

Since G is the net of P (by assumption), the cross sections of P also "intersect" G. Each xy-plane π_i (with $i \in [0, t]$) defines a cross section that intersects some faces of G on the surface of P. Before we describe how to obtain the subgraph G_i from cross section C_i, we look into how cross sections interact with faces on P. There are two options for faces intersected by an xy-plane π_i: They can be coplanar with π_i or have two of their edges parallel to π_i while the other two are perpendicular to π_i (piercing it). We only consider

---

[3] An xy-plane π_k is a plane spanned by the x- and y-axes at some constant z-coordinate k.

a discrete subset of all possible cross sections, namely those defined by xy-planes π_i with integer $i \in \{0, 1, \ldots, t\}$ and the xy-planes with coordinates half-integer above/below[4] it, namely central cross section C_i and $C_{i+\frac{1}{2}}$ and $C_{i-\frac{1}{2}}$.[5] As the realization P is aligned to the coordinate system and stands on π_0, there are three options for any face f intersected by cross section C_i (when i is an integer):

- Face f is itself coplanar to xy-plane π_i;

- One edge of f is coplanar to π_i, and either

    – "standing on" π_i (intersected by $\pi_{i+\frac{1}{2}}$); or,

    – "hanging from" π_i (also intersected by $\pi_{i-\frac{1}{2}}$).

We say that a face f of polyhedron P is a *roof face* when it is coplanar to an integral xy-plane π_i and it appears on a (solid) cube intersected by $\pi_{i-\frac{1}{2}}$ (below it). Symmetrically, a face of P is a *ceiling face* when it appears on a cube intersected by $\pi_{i+\frac{1}{2}}$ (above it). All other faces are perpendicular to the xy-plane and we call them *wall faces*. At each wall face f there are two crossing bundles – one containing the horizontal edges of f and the other containing the vertical edges of f. Given the alignment of P to the underlying coordinate system, we say that the former bundle (containing horizontal edges) extends top-to-bottom. The latter extends parallel to the half-integral xy-plane intersecting f; we refer to such bundles as *outline bundles*, as they follow the polygonal outlines of the cross sections. The half-integer indexed cross sections only intersect walls while the integer indexed cross sections collect coplanar roof/ceiling faces together with their adjacent wall faces.

Defining Pieces and Patterns. The notation presented next concerning planes, subgraphs, layers, parallel edges, bundles, and the bundle graph $\mathfrak{B}_G$ is illustrated in Figure 7.3. Considering a triple of cross sections $C_{i+\frac{1}{2}}, C_i, C_{i-\frac{1}{2}}$ of P for some integer $i \in \{0, 1, \ldots, t\}$, we can define the subgraphs $G_{i+\frac{1}{2}}, G_i$, and $G_{i-\frac{1}{2}}$ of G. Each such subgraph consists of exactly those vertices and edges of G that belong to faces on the cubes composing P intersected by each cross section respectively.[6] The half-integer indexed subgraphs consist only of wall faces – as roof and ceiling faces can only be coplanar to integer indexed xy-planes –, and G_i consists of both $G_{i+\frac{1}{2}}$ and $G_{i-\frac{1}{2}}$ as well as all roof and ceiling faces between them. Even though G and P are both connected, each such subgraph G_i can consist of multiple connected components $L_{i,1}, \ldots, L_{i,k} \in G_i$, which we refer to as *layers*. Since G_i is the graph induced by cross section C_i, we also write $L_{i,j} \in C_i$. In each layer $L_{i,j}$, we have $F_{i,j}$ as the set of faces on the intersected cubes. We use these notations for integer as well as half-integer indexed subgraphs. Notice that G_t consists of the highest roof faces P and their neighboring faces (hanging "down" from them), and G_0 consists of the lowest ceiling faces of P and their neighboring faces.

---

[4] $C_{0-\frac{1}{2}}$ and $C_{t+\frac{1}{2}}$ do not intersect P. We will discuss the triples involving either of them separately.

[5] Clearly all other non-empty cross sections are equal to one of these.

[6] Notice that "interior" cubes of P do not contain any vertices or edges and thus can be ignored.

(a)

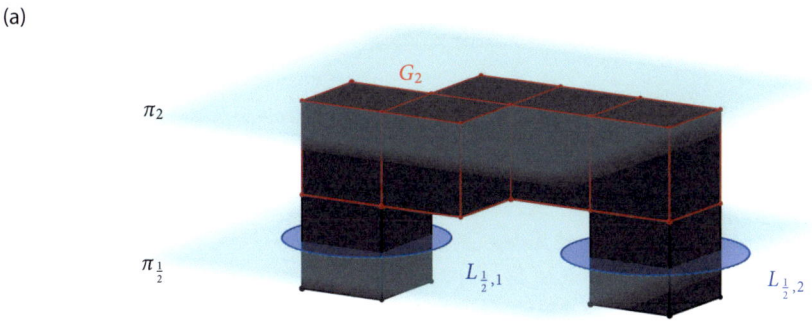

(b)

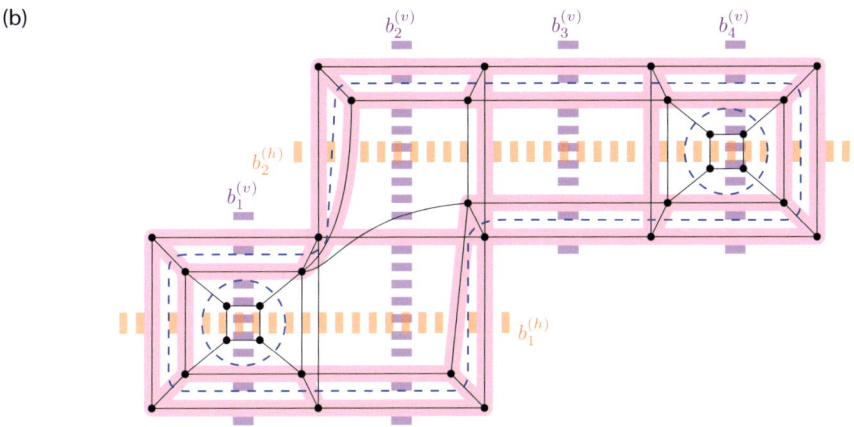

(c)

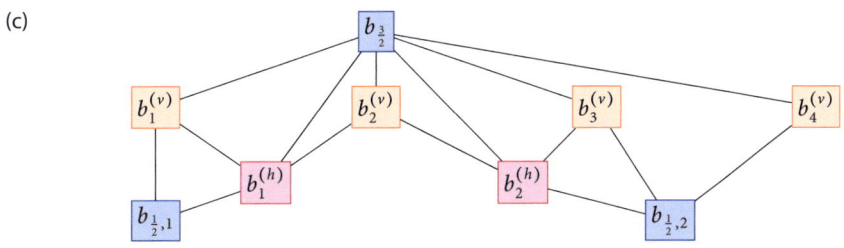

Figure 7.3: Illustration of the used notation: (a) A simple polyhedron P of height $t = 2$. Plane π_2 defines subgraph G_2 (red), plane $\pi_{\frac{1}{2}}$ defines two disconnected layers $L_{\frac{1}{2},1}$ and $L_{\frac{1}{2},2}$ (blue). The faces of P contained in π_2 are roof faces, the faces intersected by $\pi_{\frac{1}{2}}$ are wall faces. (b) The net of P. The red highlighted edges belong to G_2; the dashed lines indicate bundles: vertical bundles in purple, horizontal bundles in orange, outline bundles in blue. (c) The bundle graph $\mathfrak{B}_G$ corresponding to the net (with matching colors).

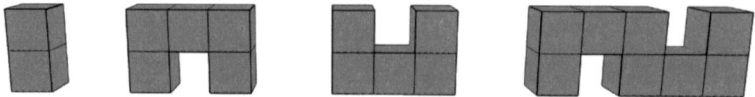

Figure 7.4: Patterns realizable by components of L_i, left to right: Straight, split, merge, and mixed.

In the following, we treat the layers of each subgraph G_i in isolation, referring to each as layer L_i – omitting the second index. We enumerate the set of five different possible patterns that we could encounter in L_i, depending on the numbers of layers in $G_{i+\frac{1}{2}}$ and $G_{i-\frac{1}{2}}$ present in L_i respectively. Patterns 2–5 (below) are illustrated in Figure 7.4.

1. In a *start* L_i only contains one layer of $G_{i-\frac{1}{2}}$, but no layers of $G_{i+\frac{1}{2}}$; symmetrically, in an *end* pattern L_i contains one layer of $G_{i+\frac{1}{2}}$, but none of $G_{i-\frac{1}{2}}$.

2. In a *straight* pattern, L_i contains exactly one layer of each of $G_{i+\frac{1}{2}}$ and $G_{i-\frac{1}{2}}$.

3. In a *split* pattern, L_i contains exactly one layer of $G_{i+\frac{1}{2}}$ but multiple of $G_{i-\frac{1}{2}}$.

4. In a *merge* pattern, L_i contains multiple layers of $G_{i+\frac{1}{2}}$ but exactly one of $G_{i-\frac{1}{2}}$.

5. In a *mixed* pattern, L_i contains multiple layers of both $G_{i+\frac{1}{2}}$ and $G_{i-\frac{1}{2}}$.

Proof Invariant. To prove Theorem 7.3, we argue that ITERATEDBUNDLECOLORING *colors* the bundles of $\mathfrak{B}_G$, i.e., the output is a partition of the bundles into three sets. This coloring will be unique (up to renaming the colors), as we will never "guess" the color of a bundle but only derive it from having two differently colored neighbors. Actually, BUNDLECOLORING itself does not exploit the geometric nature of the problem but only works on the (unique) bundle graph of G. Having the coloring of $\mathfrak{B}_G$, we can obtain an *orientation* of the faces of G, marking each of them as perpendicular to one of the three coordinate axes.

In this analysis, we will argue that ITERATEDBUNDLECOLORING would be able to "puzzle" together the results obtained from multiple executions of BUNDLECOLORING – recall the loops in the code from Algorithm 7.1. We will only consider a selected subset of all executions[7], looking for bundles that have been colored by more than one of them. When we find two executions that both color the same two bundles, we say that those executions can be *synchronized*[8], merging the color classes (renaming those of one execution). During the analysis we will oftentimes argue that two (or more) layers must have been synchronized.

While ITERATEDBUNDLECOLORING does not actually process $\mathfrak{B}_G$ top-to-bottom as we describe (since it does not know about the realization it will reconstruct), we will

---

[7] Unfortunately, we cannot identify this subset without the realization at hand. Therefore we choose exhaustive application of BUNDLECOLORING, eventually performing all required runs.

[8] This is something the algorithm finds eventually, we just argue that it has to happen.

argue using executions of BUNDLECOLORING on faces of layers in a top-down fashion. The requirements on an upper layer allowing the algorithm to color the bundles of a lower layer are stated in Lemma 7.4 below. We will not prove Lemma 7.4 directly, but instead argue for all five patterns individually.

Lemma 7.4. *Let $\mathfrak{B}_G^*$ be the output of ITERATEDBUNDLECOLORING and let $G_{i+\frac{1}{2}} = \bigcup_n L_{i+\frac{1}{2},n}$ and $G_{i-\frac{1}{2}} = \bigcup_m L_{i-\frac{1}{2},m}$ be the subgraphs of all upper and lower layers respectively. For each $i \in \{0, 1, \ldots, t\}$ we have:*

1. *If ITERATEDBUNDLECOLORING can individually orient all upper layers in $G_{i+\frac{1}{2}}$ (the bundles of each layer have been merged to form triangles in $\mathfrak{B}_G^*$),*

2. *then the outline bundles of all layers in $G_{i-\frac{1}{2}}$ would be colored by ITERATEDBUNDLE-COLORING, synchronizing them to the upper layers' outlines,*

3. *which in turn means that the remaining bundles of $G_{i-\frac{1}{2}}$ are also colored, i.e., they have been merged to form a triangle in $\mathfrak{B}_G^*$.*

Our ultimate goal is to orient the faces of G. Therefore, in this analysis we classify faces by type – either 0, 1 or 2 –, depending on the number of bundles with already fixed color class, following the notation introduced by Biedl and Genç [BG09]. Again recall that ITERATEDBUNDLECOLORING does not work with the faces of G directly; instead, we use the types of faces to discuss how much and what parts of a given layer have been processed. Property (1) of Lemma 7.4 implies that all faces in $F_{i+\frac{1}{2}}$ are of type 2; we will use this property as an *invariant* that we assume holds for all upper layers of any pattern.

Next, observe that Lemma 7.4 indeed implies Theorem 7.3: Property (1) trivially holds for $i = t$. Moreover, for each layer within any integer-indexed cross section C_i, the bundles form a triangle in $\mathfrak{B}_G^*$, by Properties 2 and 3. Thus, as our net G is connected, it must be the case that $\mathfrak{B}_G^*$ is a triangle, and as such the orientation of every face of G has been determined.

7.3.2 Lemmata for the Patterns

In the following, we fix a polyhedron P that realizes G, and consider the patterns for the layer L_i of P, establishing a lemma similar to Lemma 7.4 for each. For now, we do not consider layers with *flat holes*: Layer $L_{i-\frac{1}{2}}$ has a flat hole, when there are two (or more) lower layers $L_{i-\frac{1}{2},a}$ and $L_{i-\frac{1}{2},b}$ with outlines in $\pi_{i-\frac{1}{2}}$, such that both lower layers are connected using roof and/or ceiling faces from L_i and such that the outline of one layer is contained in the outline of the other. The most simple flat hole is shown in Figure 7.2 (a) on page 117: It has three cross sections $C_1, C_{\frac{1}{2}}$, and C_0. The middle cross section $C_{\frac{1}{2}}$ shows two outlines – one 3×3 units large, the other 1×1 (in purple) – with the larger containing the smaller and both of them connected by roof faces from C_1 (e.g. in blue). Notice that this does not imply restricting ourselves to genus-0 cases for now, as holes

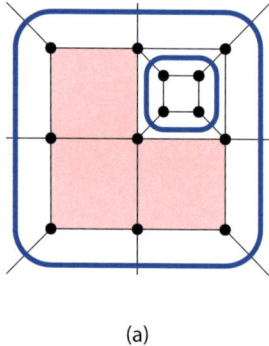

(a)
(b)

Figure 7.5: Illustration of a patch: (a) The subgraph of the net of a straight pattern; faces of the patch are drawn in red, bounding bundles in blue. (b) The same patch on the surface of the realization, forming a start pattern (with matching colors).

in planes perpendicular to xy-plane (and some other, more complicated cases) are still possible and will be handled implicitly.

To complete the analysis, we later introduce Lemma 7.12, describing how ITERAT-EDBUNDLECOLORING establishes Property 2 of Lemma 7.4 in the presence of flat holes. After coloring all bundles of a layer (Lemma 7.4 (3)), processed flat holes "disappear" from the bundle graph as all outside-inside outline pairs are colored the same and thus the bundles are merged in $\mathfrak{B}_G^*$.

Patches and Patch Spreading. In the following proofs we use the concepts of *patches* and *patch spreading*. A *patch* is a maximal connected set of faces of some subgraph G_i, defined by a "boundary"[9] of bundles with the same color: For each face f of a patch, the two bundles $b^{(1)}$ and $b^{(2)}$ crossing in f only extend across other faces of the same patch until eventually crossing one of these boundary bundles on either side. (As we will pick the patches to be coplanar to some plane π_i, these boundary bundles will in fact be outline bundles of upper and lower layers.) An illustration can be found in Figure 7.5 – even though we also provide a geometric interpretation there, the concept of patches is defined only on the faces of G and therefore implicitly present in the bundle graph.

Lemma 7.5 (Patch Spreading). *If the boundary bundles of a patch have the same color and the patch contains a face of type 2, then all faces in that patch are type 2.*

Proof. Neighboring faces share one bundle and patches are connected, thus the following argument can be applied exhaustively, spreading over all faces of the patch: If a patch

---

[9] The boundary composed of outline bundles relates to the *bands* introduced by Biedl and Genç [BG09] In the case of hole-free start- and end patterns, the single outline bundle is indeed a band.

has some type 2 face, all of its neighbors are at least type 1 (from the shared bundle); since those neighboring faces have another bundle crossing at least one boundary bundle, those neighbors are actually type 2. □

We will argue about establishing Property 3 of Lemma 7.4 using patch spreading.[10] Consider a single face of a patch – by definition, its two bundles cross each other and also cross bundles of the third color class.

Since we assume that there are no flat holes (for now), in the start, there is only one boundary outline bundle to be crossed by the bundles of patch faces. End pattern seem to be more complex since there can possibly be multiple upper layers that are crossed by bundles from faces of the patch. As those upper layers will have been lower layers of some other patterns, they are already synchronized and can be treated as one bundle. We can then either choose the color for the bundle bounding the patch (uncolored lower layer of a start pattern) or have it fixed and given (colored upper layer of an end pattern). In most of our cases however, we first ensure that Property 2 of Lemma 7.4 holds; thus, even when crossing different outline bundles, we still get the same coloring information from all of them.

With patch spreading established, we can now prove the first pattern-lemma.

Lemma 7.6 (Start- and End Pattern). *Let $L_{i+\frac{1}{2}}$ and $L_{i-\frac{1}{2}}$ be the upper and lower layer in L_i respectively.*

(a) *If $L_{i+\frac{1}{2}} = \varnothing$, an orientation of all faces in L_i can be obtained by running BUNDLE-COLORING starting on any roof face.*

(b) *If $L_{i-\frac{1}{2}} = \varnothing$ and $L_{i+\frac{1}{2}}$ is oriented, the ceiling faces in L_i can be oriented to be consistent with the coloring of $L_{i+\frac{1}{2}}$.*

Proof. For case (a), picking any roof face f coplanar to π_i yields two perpendicular bundles $b^{(1)}, b^{(2)}$. Having a realization, we know that both cross the lower outline bundle $b_{i-\frac{1}{2}}$, hence we have a patch containing f that is only bounded by $b_{i-\frac{1}{2}}$ (as L_i is connected and $L_{i+\frac{1}{2}}$ is empty). Running BUNDLECOLORING on f, we pick two colors for $b^{(1)}$ and $b^{(2)}$, making f type 2 and get that $b_{i-\frac{1}{2}}$ must be of the third color. This allows patch spreading, coloring all bundles in L_i, making all patch faces type 2. With all other bundles known, all faces of $b_{i-\frac{1}{2}}$ must also be type 2.

For case (b), consider the outline bundle $b_{i+\frac{1}{2}}$ of $L_{i+\frac{1}{2}}$. At any convex corner of this outline, there is a ceiling face f; the two bundles of f extend onto upper wall faces. By the invariant, those bundles are colored, making the corner face type 2. With $b_{i+\frac{1}{2}}$ also colored, applying patch spreading (Lemma 7.5). □

In each realizable instance, there is at least one start pattern – namely at the face defining π_t – initializing the propagation process of Lemma 7.4 and one end pattern – the face(s) coplanar to π_0.

---

[10] Again, this is not what the algorithm actually does, but a way to argue about the output.

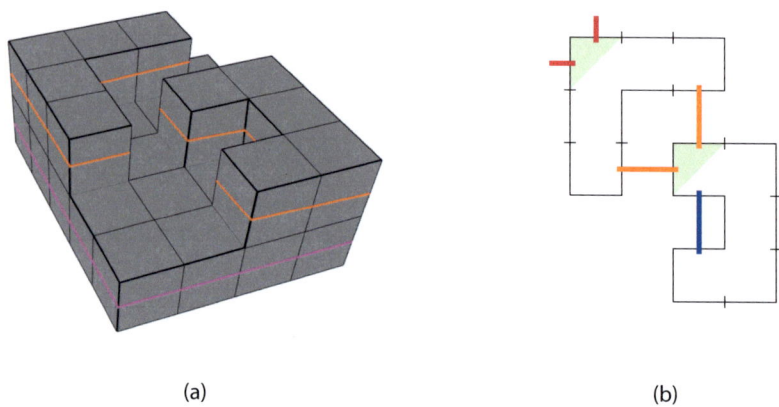

(a)   (b)

Figure 7.6: Geometric intersection and bundle directions: (a) A layer creating a merge pattern (two upper layers, one lower layer); the intersections of lower and upper layers together with their respective xy-planes in purple and orange respectively. (b) The geometric intersection of the layer from (a) yields two curves (segments indicated by ticks). The blue bundle is fake, it is northern and southern for the same layer; The two north-western corners are indicated by the green region: The corner at the orange bundles is blocked, that at the purple bundles is unblocked.

Geometric Intersection and North-Western Corners. The next rather simple pattern is the straight pattern. In this pattern we encounter an upper layer $L_{i+\frac{1}{2}}$ providing a prescribed coloring that the remaining bundles of the subgraph G_i of L_i need to be synchronized to.

Since we have the realization P at hand, we can look at the upper and lower cross sections of C_i on their xy-planes $\pi_{i+\frac{1}{2}}$ and $\pi_{i-\frac{1}{2}}$. Projecting the outlines of the intersected cubes down onto a common xy-plane, each pair of overlapping layers creates one or more two-dimensional polygonal regions called *curves*. We call the set of curves obtained by projecting and intersecting the upper and lower layers of L_i the *geometric intersection* of L_i. As the realization is aligned to a coordinate system, so are these outlines. This allows us to characterize the regions as follows:

Each unit-length segment of a curve corresponds to two faces f_1, f_2 – one from the upper and one from the lower layer – coinciding with the outline of the inner-most of the two faces. At those faces, we have three bundles in total – the two outline bundles (tracing the original polygonal regions in the planes) and some bundle b' perpendicular to both. In the following, we will oftentimes identify a curve with either of the two outline bundles of the layers creating it, depending on the layer we currently consider. By construction, b' extends over both faces creating f in the intersection, b' is therefore present in the bundle graphs of both layers. Using the alignment of P within the coordinate system, we can classify the shared bundle b' as northern, eastern, southern, or western, following the facings and perpendicular coordinate axes of f_1 and f_2 on P. By this definition, the bundle b' can be the shared bundle of multiple unit-length segments of the same curve[11], and thus be northern and southern (or eastern and western) for that same curve at the same time. In that case, we call the bundle *fake* northern and southern (or fake eastern and western). In the following, we will focus on non-fake northern and western bundles. The example found in Figure 7.6 shows these notions as well as the notation below.

On these curves, we now look for *north-western corners*. A *corner* is a convex 90° turn in the curve. It is *north-western*, if one of its segments is defined by a face with a northern bundle, the other segment is defined by a face with a western bundle, and neither of the two bundles is fake – the green areas in Figure 7.6 (b) depict north-western corners. We say that a northern (or western) bundle is *blocked* when there is a different curve for which it is southern (or eastern) – consider the orange bundle from Figure 7.6 (b). In that case, the blocked bundle traverses some wall face of another curve. Similarly, a corner is *blocked* if at least one of its bundles is blocked. It is easy to see that every curve of a geometric intersection always has at least one north-western corner, and that there always is some curve that has an unblocked north-western corner.

Lemma 7.7 (Straight Pattern)**.** *Let $L_i \in C_i$ be a straight pattern with upper and lower layers $L_{i+\frac{1}{2}} \in C_{i+\frac{1}{2}}$ and $L_{i-\frac{1}{2}} \in C_{i-\frac{1}{2}}$, respectively. Given an orientation of $L_{i+\frac{1}{2}}$, the remaining parts of L_i can be synchronized to it.*

Proof. We first establish how the color of the outline bundle $b_{i-\frac{1}{2}}$ of $L_{i-\frac{1}{2}}$ can be derived. This allows us to argue using patch spreading on the remaining roof and ceiling faces.

Let $b_{i+\frac{1}{2}}$ and $b_{i-\frac{1}{2}}$ be the outline bundles of $L_{i+\frac{1}{2}}$ and $L_{i-\frac{1}{2}}$ respectively. Consider the geometric intersection of the two outlines. In this set, consider a curve with an unblocked north-western corner – call the curve of this corner p. By Lemma 7.4 (1), the bundles defining the north-western corner of p are colored. They both cross $b_{i+\frac{1}{2}}$ as well as $b_{i-\frac{1}{2}}$, synchronizing them and establishing 2. This makes all faces of $L_{i-\frac{1}{2}}$ of type 1.

If there are patches of roof or ceiling faces in L_i, there must also be at least one face f neighboring a face of $b_{i+\frac{1}{2}}$, sharing an edge with a wall face of the upper layer and hence also sharing a bundle $b^{(1)}$. Let the other bundle of f be $b^{(2)}$ – it crosses colored bundle $b^{(1)}$ (at f) and at least one of the outline bundles making f a type 2 face for that patch, enabling patch spreading. Repeat this until all patches are oriented. □

Peeling Sequence. In the following lemma for the merge pattern, we will encounter a the geometric intersection containing multiple curves.

To synchronize all outlines, we define the *peeling sequence* of the curves of a geometric intersection using the following set of instructions: As long as there are unprocessed curves, take a curve with an unblocked north-western corner, add it to the sequence and

---

[11] For example, consider a bundle going between the ends of a "C"-shape, like the blue bundle from Figure 7.6 (b).

mark it as processed. Update the other corners by marking the northern and western bundles blocked by processed curves as not blocked (possibly marking some north-western corners as unblocked).[12]

Lemma 7.8 (Merge Pattern). *Let $L_i \in C_i$ be a merge pattern with $L_{i+\frac{1}{2},1}, \ldots, L_{i+\frac{1}{2},k} \in C_{i+\frac{1}{2}}$ and $L_{i-\frac{1}{2}} \in C_{i-\frac{1}{2}}$. Given locally consistent orientations for all layers in $C_{i+\frac{1}{2}}$, (i.e., in each $L_{i+\frac{1}{2},j}$ the bundles have been merged to form a triangle in $\mathcal{B}_G^*$ – Lemma 7.4 (1)) the remaining parts of L_i can be oriented consistently (as in Lemma 7.4 (2) and (3)).*

Proof. We pick one of the upper layers as a reference and first argue how BUNDLECOLOR-ING synchronizes the other outline bundles to it. We argue how the roof and ceiling faces between them become oriented by patch spreading. With these orientations in place, we then synchronize the color classes of the remaining layers into the reference layer.

Consider the geometric intersection of all upper layers with lower layer $L_{i-\frac{1}{2}}$; for every upper layer, we get some curve corresponding to it. Let $L_{i+\frac{1}{2},1}$ be the layer of the first element of the peeling sequence for that geometric intersection. This curve has a north-western corner with two unblocked bundles, i.e., they both cross $b_{i-\frac{1}{2}}$ and $b_{i+\frac{1}{2},1}$ only extending over roof or ceiling faces in between. Having bundles of two different color classes identified in both layers, we synchronize $L_{i-\frac{1}{2}}$ and $L_{i+\frac{1}{2},1}$.

Following the peeling sequence, we synchronize all other outline bundles as follows: Consider the representative north-western corner c of the current element in the sequence belonging to upper layer $L_{i+\frac{1}{2},j}$. The bundles at c form a triangle with outline bundle $b_{i+\frac{1}{2},j}$. If both bundles at c are not blocked, then they cross the lower layer's outline bundle $b_{i-\frac{1}{2}}$. Thus $b_{i+\frac{1}{2},j}$ gets synchronized to the lower layer's outline (like $b_{i+\frac{1}{2},1}$ was). If either (or both) of the bundles is blocked, it must be blocked by some curve to the north (or west respectively), crossing that upper layer's outline bundle. As that curve needs to be earlier in the sequence (by being more north and/or west), its outline bundle is already synchronized, hence propagating the same coloring information as if crossing $b_{i-\frac{1}{2}}$. This way, ITERATEDBUNDLECOLORING can synchronize all outline bundles of L_i.

With all upper and lower layers synchronized, the missing patches of roof and ceiling faces can be synchronized using patch spreading like above. $\qquad\square$

Hooking Curves Together. As opposed to the merge pattern above, the multiple lower layers of the split pattern are not covered by the invariant of Lemma 7.4 (1) – that is, identifying two perpendicular bundles crossing a curve does not imply synchronizing the layers they are obtained from, as the bundles of those layers are not grouped into three color classes yet.

To overcome these problems, we use a similar strategy to that of Lemma 7.8: We look for the north-western corners (with bundles $b^{(h)}$ and $b^{(v)}$) in sequence, deriving the color for each corresponding outline bundle from the upper layer's outline or the curves

---

[12] For the purpose of defining this sequence, use the following intuition: Imagine removing the picked curve from the geometric intersection, "peeling it away" to make other curves accessible. We will later introduce Lemma 7.9 to argue how this peeling relates to ITERATEDBUNDLECOLORING.

blocking it. In the following discussion we look for a face f "near" the corner; that is, the two faces creating corner c and face f share exactly one vertex in G (they are diagonally opposite on P). To obtain a unique valid coloring despite having blocking curves, we argue having an execution of BUNDLECOLORING "nearby" and fixing the colors of two additional bundles $b^{(1)}$ and $b^{(2)}$, allowing us to explore the crossing patterns of those six bundles (including the two outlines). A schematized geometric intersection (with faces, corner, and bundles) is shown in Figure 7.7 (a). The subgraph characterizing the bundles for this configuration is shown in Figure 7.7 (b).

We demonstrate how the result of a single execution of BUNDLECOLORING on f *hooks* the outline bundles of the curves to each other. Using the two crossing bundles starting in f, we can derive that the outline bundles $b_{i,1}$ and $b_{i,2}$ have to be in the same color class, even if neither outline has been synchronized to an upper outline bundle yet.

It is also possible that there is no face f diagonally opposing c; consider that realization P might be missing a cube in the upper layer. In that case, we technically cannot argue using Lemma 7.9. If there is no cube on the upper layer to have face f on its surface, this missing cube certifies that there are two bundles coming down from the upper layer, one being coplanar to $b^{(1)}$ and the other coplanar to $b^{(2)}$. As those bundles are colored differently by the invariant and cross an upper outline bundle, we can use these colors instead of those obtained by the "missing" execution. Due to the merging bundles of the same color in $\mathfrak{B}_G$, we can still assume to have the local structure in the bundle graph shown in Figure 7.7 (d).

Lemma 7.9 (Hooking layers). *Let C be the set of curves of the geometric intersection for some layer L_i. Suppose C is being processed using a peeling sequence and let c be the north-western corner of the currently processed curve. When corner c is blocked, the curve of c can be hooked to curves processed earlier in the sequence – either by using a suitable execution of BUNDLECOLORING or by considering the structure of P nearby c.*

Proof. Let $b_{i,1}$ be the outline bundle of the currently processed curve, let c be the north-western of this curve and let $b^{(h)}$ and $b^{(v)}$ be the western (horizontal) and north-ern (vertical) bundles of c. Suppose that both $b^{(h)}$ and $b^{(v)}$ are blocked by other curves. This layout of curves is depicted in Figure 7.7 (a); the structure in the bundle graph that enables the hooking is shown in Figure 7.7 (b), with $b_{i,2}$ being the representative of all outline bundles previously synchronized.

Since bundle $b^{(h)}$ is blocked by some curve, there must be at least one face between c and the blocking curve. Let f_h be the first of those faces, neighboring the wall face defining c. Symmetrically define f_v as the first face of $b^{(v)}$. Consider the two other bundles of faces f_v and f_h, call them $b^{(1)}$ and $b^{(2)}$ respectively. By construction, these bundles cross at some face f. This face f is diagonally opposite of c on P, sharing one vertex with the wall faces composing c.

To derive that $b_{i,1}$ and $b_{i,2}$ will be placed into the same color class, we need to show two things: First, $b^{(h)}$ and $b^{(v)}$ both cross $b_{i,1}$ and some outline bundle synchronized to $b_{i,2}$; and second, that $b^{(h)}$ and $b^{(v)}$ will be placed in different color classes.

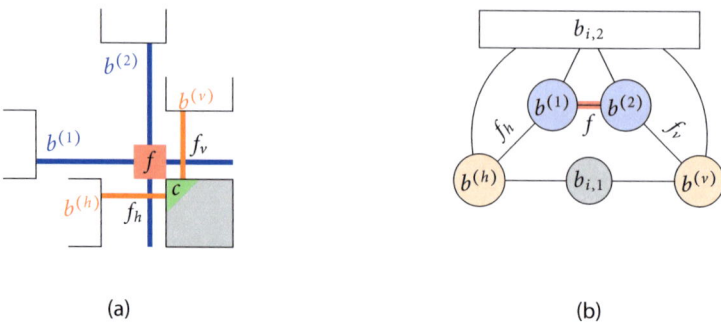

(a)   (b)

Figure 7.7: Hooking a blocked curve to the already processed parts of $\mathfrak{B}_G$, see Lemma 7.9. (a) Corner c (in green) has two (orange) bundles $b^{(h)}$ and $b^{(v)}$ that are both blocked by curves to the west and north respectively. Face f (in red) is diagonally opposite of c, the two (blue) bundles $b^{(1)}, b^{(2)}$ of f extend to the west and north respectively, eventually crossing other outline bundles (either other curves or the upper layer's outline). The pairs of bundles $b^{(h)}, b^{(2)}$ and $b^{(v)}, b^{(1)}$ cross at f_h and f_v respectively. (b) A subgraph of the bundle graph $\mathfrak{B}_G$ for the layers of (a) (with matching colors).

The first part is easy: By construction both bundles cross $b_{i,1}$. As $b^{(h)}$ extends to the west, any curve blocking it must have a north-western corner further to the west. Thus, this curve was processed before the current curve, meaning it has been synchronized to the outline bundle. If $b^{(h)}$ is not blocked, it trivially crosses the outline bundle that the current curve is supposed to be synchronized to. This argument holds symmetrically for bundle $b^{(v)}$.

For the second part, consider the execution of BUNDLECOLORING with starting face f. The two bundles $b^{(1)}$ and $b^{(2)}$ of f cross each other, and by extending the the west/north, both also cross representatives of $b_{i,2}$. As $b^{(v)}$ crosses $b^{(1)}$ and both cross some bundle synchronized to $b_{i,2}$, both bundles also need to be in different color classes. This forces $b^{(1)}$ and $b^{(h)}$ to have the same color (symmetrically for $b^{(2)}$ and $b^{(v)}$). We now have that $b^{(h)}$ and $b^{(v)}$ are in different color classes, completing the claim.

Notice the following: If any of the three faces f, f_h, f_v does not exist, we get that (at least) one of the pairs of crossing bundles directly cross the outline we are trying synchronizing to, ensuring that both bundles are in different color classes. □

Since we cannot know the correct peeling sequence while running ITERATEDBUNDLE-COLORING – the algorithm does not know the realization P –, we rely on exhaustive application. In particular, the hooking via the peeling sequence means that – assuming that the faces of the upper layers are fully oriented – some bundles can be merged by a single execution of BUNDLECOLORING, and as such, ITERATEDBUNDLECOLORING would do so. As we have discussed, we can assume that such a sequence exists for any valid positive instance. After hooking two curves, both outlines propagate the same coloring information when crossed, so we can effectively treat the layers creating them as as having a single outline bundle, hence merging them.

Lemma 7.10 (Split Pattern). *Let $L_i \in C_i$ be a split pattern with upper layer $L_{i+\frac{1}{2}} \in C_{1+\frac{1}{2}}$ and lower layers $L_{i-\frac{1}{2},1}, \dots, L_{i-\frac{1}{2},m} \in C_{1-\frac{1}{2}}$. Given an orientation of $L_{i+\frac{1}{2}}$, the remaining parts of L_i will be synchronized to it by ITERATEDBUNDLECOLORING.*

Proof. Let $b_{i+\frac{1}{2}}$ be the outline bundle of the upper layer. As before, we first establish that all lower outline bundles will be synchronized to $b_{i+\frac{1}{2}}$.

Again, consider the geometric intersection of L_i and let $L_{i-\frac{1}{2},1}$ be the lower layer contributing to the first curve in a peeling sequence. As the first element cannot be blocked, there are two bundles present in $L_{i-\frac{1}{2},1}$ that are uninterrupted. Hence, these bundles allow us to synchronize $b_{i-\frac{1}{2},1}$ and $b_{i+\frac{1}{2}}$.

We now consider the remaining elements of the peeling sequence in order, synchronizing each outline bundle to $b_{i+\frac{1}{2}}$. Let c be the north-western corner of the current curve. If c is not blocked, the northern and western bundle each cross both outlines, synchronizing them. If c is blocked, we argue using Lemma 7.9: We observe that there is a face f on which BUNDLECOLORING can be executed, such that both outlines are hooked. The hooking certifies that the outline of the curve of c is synchronized to the outlines blocking it.

This shows that all lower outlines are synchronized to the upper outline bundle $b_{i+\frac{1}{2}}$, making all wall faces of L_i at least type 1. Next, consider the patches of roof and ceiling faces. Since L_i is connected, all patches must have a face neighboring a wall face of the upper layer's outline bundle $b_{i+\frac{1}{2}}$. At that wall face, there is a bundle $b^{(1)}$ (perpendicular to $b_{i+\frac{1}{2}}$) extending at least one face of the patch; let f^* be that face, and let $b^{(1)}$ and $b^{(2)}$ be the two bundles of f^*. By construction, the bundle $b^{(1)}$ of f^* belongs to the upper layer, and the other bundle $b^{(2)}$ crosses some synchronized outline bundle. Thus, $b^{(2)}$ crosses two colored bundles and will be colored itself. This implies that f^* is of type 2. Applying patch spreading (Lemma 7.5) to that patch colors all bundles of faces in the patch of f^*. Repeating this argument until all patches are oriented concludes the proof. $\square$

To see the conceptual difference between the mixed pattern and regular split or merge patterns, consider the single central cube in the upper layer of in Figure 7.8 (a); call it $L_{i-\frac{1}{2},c}$. While the geometric intersection of the outlines is non-empty, neither of the two bundles going over $L_{i-\frac{1}{2},c}$ crosses the outline of the layer below it – both extend over ceiling faces and onto other upper layers on each side. To apply either Lemma 7.8 or Lemma 7.10, we would require all outline bundles to be in the same color class, but the alternating structures created by upper and lower layers prevent us from using either lemma.

Lemma 7.11. *Let L_i be a mixed pattern with upper layers $L_{i+\frac{1}{2},1}, \dots, L_{i+\frac{1}{2},k} \in C_{i+\frac{1}{2}}$ and lower layers $L_{i-\frac{1}{2},1}, \dots, L_{i-\frac{1}{2},\ell} \in C_{i-\frac{1}{2}}$. Given individual orientations for all upper layers in $C_{i+\frac{1}{2}}$, the all of L_i will be synchronized.*

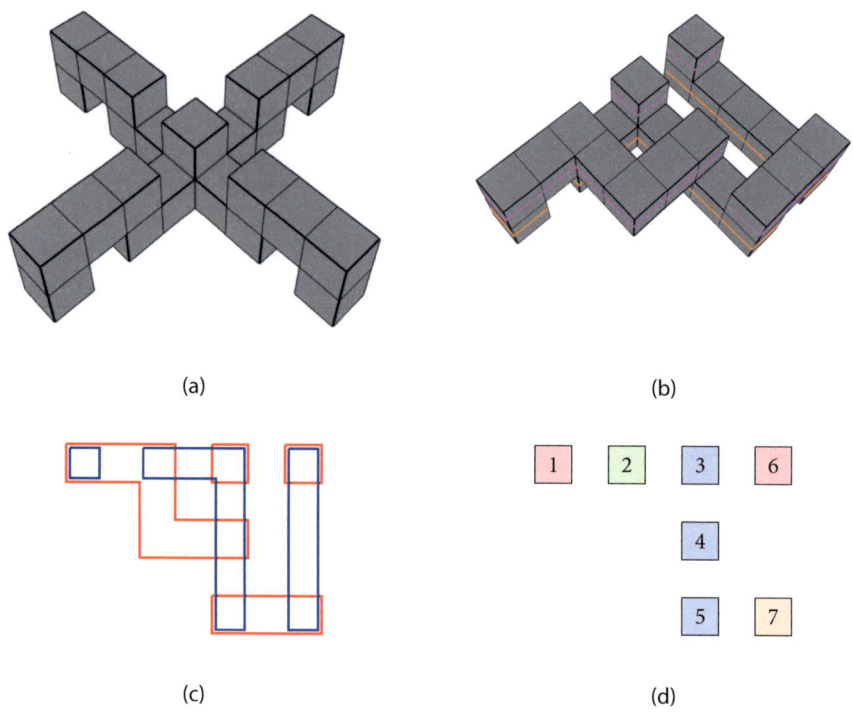

Figure 7.8: The mixed pattern: (a) The central layer $L_{i+\frac{1}{2},c}$ of this pattern has no bundle hitting the outline of the layer it stands on. (b) A layer of a polyhedron forming a mixed pattern, three upper outline bundles in purple and four lower outline bundles in orange. (c) The set of upper (in red) and lower outlines (in blue) from (b); (d) the geometric intersection of the two sets from (c) labeled from 1 to 7 by sequence of consideration, Start cases in red (1 and 6), All-Lower case in green (2), All-Upper cases in blue (3, 4 and 5), and Set-Merge case in orange (7).

Proof. In the previous patterns we could always assume that all layers of one side (upper or lower) interact with the same layer on the other side. For the mixed pattern, this is no longer the case: When synchronizing the outline bundles of two upper-lower-layer pairs, we might end up with two *subsets* of merged bundles. Since layer L_i is connected in P, so is the corresponding subgraph of the bundle graph and eventually all outlines will be merged (given the right executions of BundleColoring).

Recall that a curve of a geometric intersection is created by intersecting the outlines of two layers – an upper layer L_u overlapping a lower layer L_ℓ with outline bundles b_u and b_ℓ. A northern (or western) bundle is blocked when it starts on a face of b_u (or symmetrically b_ℓ) and extends upwards (to the left), but instead of crossing b_ℓ it crosses the outline bundle of some other curve. If a bundle starting on a face of b_u (or b_ℓ) is blocked, it must cross the outline bundle of some other upper (lower) layer; we distinguish between

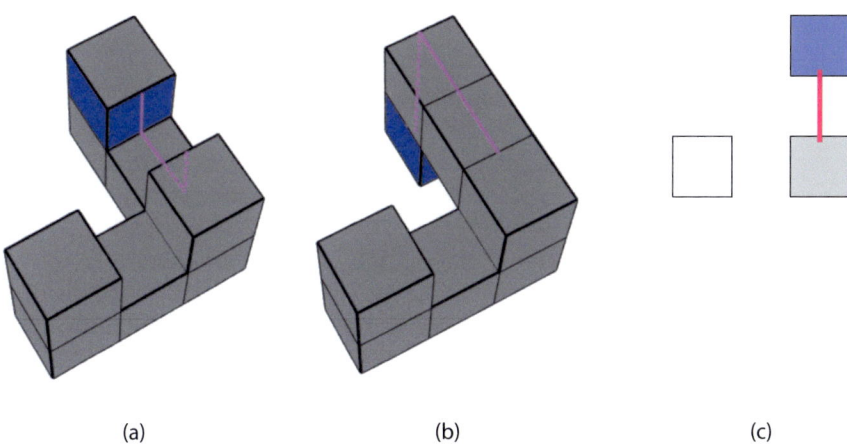

      (a)   (b)   (c)

Figure 7.9: Different blocking patterns creating the same geometric intersection. The geometric intersection shown in (c) can either be obtained from the polyhedron shown in (a) or (b). The purple bundle is blocked in both configurations: In (a) we have an upper block and in (b) we have a lower block.

these options, calling them *upper* and *lower* blocks respectively. The different blocks are illustrated in Figure 7.9.

    We now classify the curves by the bundles on their north-western corner c. If both bundles of c are unblocked, the two layers start a new subset. In all other cases, at least one of the bundles at c is blocked by some curve. Since this curve is further north (or west) in the intersection, it was processed before – the layers creating the curve already belong to some subset. With this in mind, we classify the remaining curves by the curves blocking the bundles of their corners, depending on whether all blocked bundles are blocked by curves created from layers of same subset[13] or from different subsets.

    This leads to the following four cases:

(1) *Start:* No blocks – starting a subset – similar to the straight pattern (Lemma 7.7).

(2) *All-Upper:* All blocks are upper blocks, similar to the merge pattern (Lemma 7.8).

(3) *All-Lower:* All blocks are lower blocks, similar to the split pattern (Lemma 7.10).

(4) *Set-Merge:* One block is an upper block, the other block is a lower block.

An illustration of these cases can be found in Figure 7.8 (b) to (d). In the following, we consider each case, arguing how to synchronize the outline bundles. This will later allow us to use patch spreading on the roof and ceiling faces on L_i.

---

[13] This naturally includes the case when only one bundle is blocked.

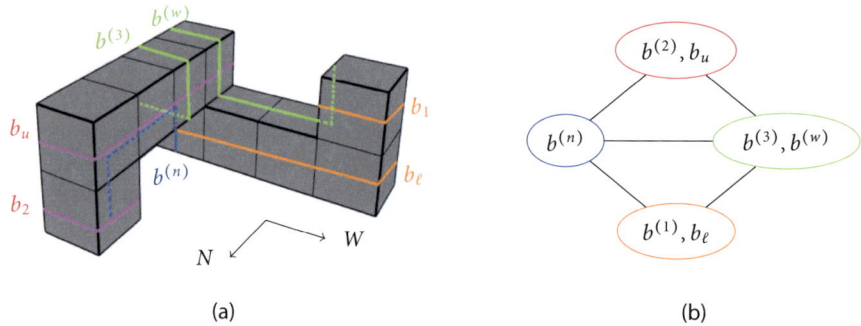

$$(a) \qquad\qquad (b)$$

Figure 7.10: Illustration of the Set-Merge Case (4). (a) A simple polyhedron showing a Set-Merge Case (rotated such that the viewpoint is the north-western side, compass rose included). The colored bundles indicate synchronization: Purple and green for the outlines of the two subsets respectively, green for the western bundles of the upper layer, blue for the (supposed) northern bundle of the lower layer. (b) The part of the bundle graph showing how the four colors interact: Green-purple and green-orange edge from $b^{(w)}$ crossing b_u and b_1 on a wall face respectively; blue-purple and blue-orange edge from $b^{(n)}$ crossing b_ℓ and b_2 on a wall face respectively; blue-green edge from $b^{(3)}$ crossing $b^{(n)}$ on the ceiling face neighboring the corner.

For Case (1), we trivially synchronize the two outline bundles – both bundles at corner c are unblocked.

The All-Upper Case (2) can only occur when synchronizing the outline of a thus far unprocessed upper layer into a subset. The new layer and all layers of blocking curves share the same lower layer L_ℓ (otherwise, it would not have been an upper blocking). Since the blocking curves are more northern and/or western, they have been synchronized to b_ℓ before. As the bundles of L_u are 3-colored (by the invariant of Lemma 7.4 (1)), the outline bundle b_u can be synchronized.

Symmetrically, the All-Lower Case (3) can only occur when synchronizing an unknown lower layer into a subset. All blocking layers share the same upper layer L_u and the outlines of the layers creating the blocking curves are already synchronized to b_u. This allows us to hook the new layer's outline b_ℓ to the outlines of the blocking curves (using Lemma 7.9), synchronizing them.

For the Set-Merge Case (4), we are given a curve with corner c an upper and a lower block. Assume that the upper block is to the west and created by layer L_1 (with outline bundle b_1) overlapping L_ℓ and that the lower block is to the north and created by layer L_2 (with outline bundle b_2) overlapping L_u. This configuration is illustrated in Figure 7.10 (a). Let c_n and c_w be the northern and western curve respectively. Assume that A, B, and C are the three color classes of L_u from the invariant. From processing curve c_n we know that b_2 and b_u have the same color, say this color is A; symmetrically we know from c_w that b_1 and b_ℓ have the same color, say D as it is yet to be synchronized. To establish that b_u and b_ℓ have the same color – merging A and D, synchronizing all four outlines –, we look at the two bundles at c, namely $b^{(w)}$ and $b^{(n)}$. From the pattern we have that one of the two bundles is synchronized to the color classes of L_u (by extending over a wall face of L_u, participating in the upper block) whereas the other is not; w.l.o.g. assume that this bundle is $b^{(n)}$ and that its color is B. Since $b^{(n)}$ crosses b_1, we have that $D \neq B$. To argue that the outline bundles get synchronized – that is, $D = A$ – we now need to show that the color of $b^{(w)}$ needs to be C, as $b^{(w)}$ crosses b_ℓ. In Figure 7.10 (b), we give a corresponding bundle graph with A being purple, B being green, C being blue, and D being orange.

Consider the first ceiling face f_1 that $b^{(w)}$ extends over after the wall face at which it crossed b_ℓ. Let bundle $b^{(3)}$ be the other bundle of f. Symmetrically, let f_2 be the first roof face of bundle $b^{(n)}$. Figure 7.10 (a) suggests that $b^{(3)}$ must "wrap up" – that is, $b^{(3)}$ extends onto a wall face of L_u, making it part of L_u and thus colored by the invariant. To show that this wrapping up indeed occurs in a valid realization, consider the unit square x diagonally opposing corner c in the geometric intersection of cross section π_i. Considering how P is built from unit cubes, first observe that x is empty space in the geometric intersection – that is, x cannot be another corner, created by two cubes (one intersected in $\pi_{i+\frac{1}{2}}$ and one cube intersected in $\pi_{i-\frac{1}{2}}$). Wrapping up could still be prevented from happening by a single cube in either $\pi_{i-\frac{1}{2}}$ or $\pi_{i+\frac{1}{2}}$ – possibly making $b^{(3)}$ extend over another wall or ceiling face respectively. Neither case is possible: The upper cube would share the same edge with face f_2 and a wall face of L_ℓ on P but not in G, making P degenerate; symmetrically, the lower cube would share the same edge with f_1 and a wall face of L_u. Therefore, x must indeed mark empty space in P, allowing $b^{(3)}$ to wrap around. The construction of $b^{(3)}$ implies that it has the same color as $b^{(w)}$, therefore we get that the color of $b^{(n)}$ is neither A nor B and thus C. We now have that the bundles of color class D are crossed by bundles of colors B and C, enforcing the synchronization of classes A and D.

This concludes the enumeration of all possible patterns involving upper and lower blocks, thus arguing that ITERATEDBUNDLECOLORING eventually synchronizes all outline bundles in the mixed pattern. With all outlines synchronized, patch spreading synchronizes the bundles on roof and ceiling faces of π_i. Since each patch must share a bundle with some upper layer, all layers will get synchronized. □

The following – and final – lemma reconsiders all patterns above in the presence of flat holes. Recall that a flat hole is a pair of outline bundles such that they both belong to a single layer of the upper half-integral cross section – creating an inside-outside outline pair. Notice that flat holes (and the bundles creating them) are not immediately visible in either the supposed net G or the bundle graph $\mathfrak{B}_G$. Nevertheless, given a realization P that contains flat holes, we can argue how ITERATEDBUNDLECOLORING discovers the colors creating the orientations of the faces on P.

Any such inside-outside outline bundle pair corresponds to the existence of one hole, but one outer outline might contain several independent inner outlines and thus be part of multiple pairs. In addition to transmitting coloring information downwards from

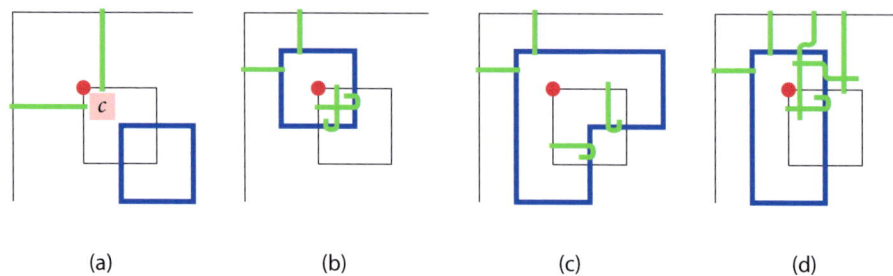

(a)   (b)   (c)   (d)

Figure 7.11: Patterns for synchronizing the inner outline of a hole to the outside outline using the upper layer (outline in blue), depending on the northern and western bundles provided by the upper layer. (a) Corner c (purple dot) is not covered by the upper layer – the bundles at c synchronize b_i to curves more northern/western. (b and c) c is covered by the upper layer, but the upper layer has two bundles of different colors wrapping down and crossing b_i, synchronizing both outlines. Notice that in (c), the northern bundle does not wrap around but follows the wall faces. (d) The corner is covered, but there is no northern bundle from the upper layer going into the hole, so synchronizing a (northern) bundle from a roof face is required.

the (upper) layers of $C_{i+\frac{1}{2}}$ to the (lower) layers of $C_{i-\frac{1}{2}}$, we now also have to consider transmitting "sideways" from one hole outline to the other.

Since we aim to process P using the patterns presented above, we have to distinguish between holes in starting patterns and holes in lower layers. We do not need to worry about holes in the upper layers of other patterns, because by the invariant, all bundles in these upper layers were lower layers of the previous step and subsequently have been merged and virtually disappear from the contemporary bundle graph $\mathcal{B}_G^*$.

Lemma 7.12 (Hole-Layer Lemma). *Let L_i be one of the patterns above.*

    (a) *If L_i is a start pattern and lower layer $L_{i-\frac{1}{2}}$ contains holes, there is a set of executions of BUNDLECOLORING that merges all outline bundles in $L_{i-\frac{1}{2}}$ into the same class.*

    (b) *If L_i is an end pattern and upper layer $L_{i+\frac{1}{2}}$ contains holes, there is a set of executions of BUNDLECOLORING on ceiling faces that merges all outline bundles in $L_{i-\frac{1}{2}}$ into the same class.*

    (c) *If L_i is not a start or end pattern pattern and lower layer $L_{i-\frac{1}{2}}$ contains holes, all bundles in $L_{i-\frac{1}{2}}$ corresponding to inner outlines either get hooked to the outer outline bundle containing it, or synchronized to the outline bundle of some upper layer.*

Proof. Our only concern here is to argue about how the outline bundles get synchronized using specific executions of BUNDLECOLORING. Once we can safely assume that all outline bundles are synchronized, patch spreading carries over to patches with holes. In fact, the net G (and also the bundle graph $\mathfrak{B}_G$) of a hole in a lower layer is indistinguishable from an upper layer "standing" at the same spot.[14]

Regarding (a): Among the outlines of $L_{i-\frac{1}{2}}$, consider a (convex) north-western corner of the outer outline. Executing BundleColoring at the roof face of this corner initially fixes the three color classes that all inner outlines will to be synchronized to, argued as follows. We now imagine that the inner outlines are processed using a peeling sequence – despite not having a geometric intersection here, the (empty) corners of inner outlines will have northern and western bundles that are either blocked by the outlines of other holes or reach the outside outline, allowing us to define a similar peeling sequence.[15] This allows us to apply Lemma 7.9, hook all holes together and eventually synchronize all of them to the outside outline using executions of BundleColoring on various roof faces.

Regarding (b): The outline bundles of holes in the upper layer of an end pattern have been part of lower layers of the previous layer. Thus, each hole provides its own 3-coloring of its bundles. Since the outside outline containing all holes is also colored, we are in a setting similar to the Merge Pattern (Lemma 7.8): Considering a peeling sequence on the holes, we can hook the inner outlines to the outer outline and each other using executions of BundleColoring on ceiling faces. Eventually all outlines will be synchronized, leaving one patch of ceiling faces that can be processed using patch spreading (Lemma 7.5).

Regarding (c): For any other pattern, we can have flat holes in some lower layer $L_{i-\frac{1}{2}}$. In that case, we synchronize the inner outlines to either the lower outline $b_{i-\frac{1}{2}}$ (containing it) or the outline bundle of some (upper) layer from $L_{i+\frac{1}{2}}$. We do this by arguing that we can either add it to a peeling sequence for the upper layers or that we can synchronize the outline bundle b_i to that of the upper layer covering it. In either case, we require that the outline bundles of the upper layers all get synchronized to the lower layers outline bundle by following the reasoning provided in the lemmata for the individual patterns.[16] For a flat hole with inner bundle b_i, there are four cases depending on the "shape" of the ceiling over c[17] provided by upper layers. The four possible shapes are depicted in Figure 7.11:

(1) None of the bundles at c are part of the upper layer, leaving c *uncovered* (Figure 7.11 (a)),

(2) corner c is covered and there are two bundles of different color from the upper layer both crossing b_i, with two different options shown in (Figure 7.11 (b) and (c)), or

(3) there are only bundles of one color from the upper layer crossing b_i (Figure 7.11 (d)).

---

[14] The gadgets used in the $\mathcal{NP}$-hardness reduction by Biedl and Genç [BG08] exploit this.

[15] Recall that the definition of a north-western corner disallows using fake northern (or western) bundles.

[16] Here we have the additional complication that holes can actually block northern and western bundles. To do so, however, such a hole must be more to the north/west and thus have been synchronized before.

[17] Remember that c is the corner of a hole, and thus, that c is empty space.

We now discuss the cases individually, arguing how ITERATEDBUNDLECOLORING synchronizes the inner outline of the hole to the upper layer's outline.

(1) In the uncovered case (Figure 7.11 (a)), there is no upper layer covering c. In this case, we can handle synchronization of b_i as if it were an upper layer – exploiting the fact that inner lower layers and upper layers are indistinguishable in $\mathfrak{B}_G$. Treating it like an upper layer, there is some peeling sequence containing it. By the time that b_i is processed, we know that the two bundles at c – northern bundle $b^{(n)}$ and western bundle $b^{(w)}$ – each either crosses the lower layer's outline or is blocked by some other upper curve. By choice of the sequence, we know that all those bundles are already synchronized. Therefore, b_i can be synchronized using $b^{(n)}$ and $b^{(w)}$.

(2) In the two-sided cases (Figure 7.11 (b) and (c)), there is a ceiling face over c and the outline of the layer covering c has a corner over the hole defined by inner outline b_i. Let that corner in the covering layer be c'. In either case – c' being convex or reflex – there are two bundles of different colors from an upper layer wrapping around onto the ceiling over c. These bundles eventually cross b_i, synchronizing it to the upper layers outline.

(3) In the one-sided case (Figure 7.11 (d)), corner c is covered by a ceiling face but we only have bundles from one direction (w.l.o.g. say west) from the upper layer but not from the other direction (say north). Since the upper layer covers c, it must have a north-western corner further north and west – hence, that layer's outline bundle is already synchronized to the outside outline bundle. To find a northern bundle crossing b_i, consider the face neighboring both outline bundles – b_i and the upper layer's outline. This face must exist as the hole is not completely covered by the upper layer. At that face, there are two bundles: Bundle $b^{(1)}$ traverses a neighboring wall face of the upper layer, marking it as western. The other bundle $b^{(2)}$ extends towards the north, eventually crossing some other outline (that has already been processed by choice of sequence). This marks $b^{(2)}$ as northern; as it wraps around and down onto a wall face in b_i, this synchronizes the outline bundles.

□

With all of the pattern lemmata in place, we can now prove Lemma 7.4 and subsequently Theorem 7.3.

Proof of Lemma 7.4. Assuming that a realization exists, there is at least one triple of bundles forming a guaranteed set of merged bundles for $\mathfrak{B}_G^*$ – namely that of the face defining π_t. Therefore we are guaranteed that there is at least one starting pattern, that can be colored and subsequently have its bundles merged according to Lemma 7.6 (a) (or Lemma 7.12 (a), respectively), establishing the condition of (1), referred to as the invariant.

Discovering and coloring more outline bundles for condition (2) – while traversing the realization downwards – we can argue that all encountered bundles can be merged using only the coloring information locally available in the triples of layer subgraphs. This is done by either Lemma 7.8, Lemma 7.10 or Lemma 7.11. In each lemma, we first argue how IteratedBundleColoring preserves the invariant by being able to merge the bundle-triangles of multiple upper layers (stemming from different starting patterns). Then, all remaining outline bundles of the current layer will be synchronized to the upper outlines: Outer outline bundles get handled directly in the individual lemmata; inner outlines of flat holes are implicitly added to the synchronization process – this is argued in Lemma 7.12.

Finally, in each layer, all missing bundles of roof and ceiling faces get colored using patch spreading, arguing that condition (3) holds, concluding the proof. □

Proof of Theorem 7.3. In this section, we considered pairs of xy-parallel cross sections at unit distance, exhaustively enumerating all possible patterns that can be encountered in a valid realization. When our input graph is indeed the net of a polyhedron, the invariant established in Lemma 7.4 states that IteratedBundleColoring is able to orient all faces on that polyhedron by coloring the bundles in $\mathfrak{B}$.

For each of the patterns, we demonstrate that there is a finite number of single executions of BundleColoring that together merge the color classes into a triangle. Each merge is unambiguous, resulting in IteratedBundleColoring producing a stable and unique coloring if one exists. Exhaustive execution as done by IteratedBundleColoring implies that the algorithm does not stop attempting to make progress using BundleColoring unless no starting face would merge bundles. Hence, if IteratedBundleColoring stops at some point where $\mathfrak{B}_G^*$ is not a triangle, some part of the graph could not be realized using any of the patterns, and we have an non-realizable instance. □

7.4 Conclusion

In this chapter, we introduced the IteratedBundleColoring algorithm that uses the original BundleColoring algorithm by Biedl and Genç [BG09] as a subroutine. We showed that exhaustive and repeated execution of the original algorithm can be used to report all edges of the graph for which the dihedral angles must be 180° in any orthogonal polyhedral surface that realizes this graph and facial angles. This implies that any orthogonal polyhedral surface – even those of arbitrary genus – arising from an orthogonal polyhedron whose graph is connected are rigid – that is, the net and the facial angles together with the edge lengths determine the dihedral angles.

Our algorithm has a total runtime in $O(m^3)$, but our focus in this chapter was put onto proving correctness of IteratedBundleColoring. We believe that the runtime analysis can be improved by considering the sequence of BundleColoring executions more carefully. The question of non-orthogonal polyhedral surfaces also remains open.

Chapter 8

Conclusion

The contents of this book are dedicated to problems regarding constrained graph layouts. We picked two general ideas – convex drawings and grid drawings –, looking into two specialized problems for each idea. This division gave birth to this book's two-part structure. We now briefly recall the main results of each chapter in order, highlighting open questions and interesting opportunities for future research.

Part One: Drawing Vertices on a Common Outer Face

Beyond Outerplanarity. The main contribution of Chapter 3 was looking into the combination of graph properties – outerplanarity and k-(quasi-)planarity – obtaining convex graph drawings with two different restrictions: limiting the number of crossings per edge or limiting the size of any sets of pairwise crossing edges.

We showed that outer k-planar graphs are $(\lfloor \sqrt{4k+1} \rfloor + 1)$-degenerate, thus obtaining a coloring bound on these graphs. We have also shown that they have small balanced separators (of size $2k + 3$), allowing outer k-planarity to be tested in quasi-polynomial time.

By giving graphs of either class that are not member of the other, we have shown that the classes of outer k-quasi-planar graphs and planar graphs are incomparable. We have also proven that all maximal outer k-quasi-planar graphs are also maximum – for any outer k-quasi-planar graph, there is a way to add missing edges until the upper bound of $2(k-1)n + \binom{2k-1}{2}$ is reached.

We also gave a linear-time algorithm to recognize full and closed outer k-planar graphs. It directly follows from Courcelle's Theorem [Cou90] (by expressing closed outer k-planarity in Monadic Second-Order Logic) and the fact that outer k-planar graphs have bounded treewidth.

To this end, the following questions remain open. Since Auer et al. [ABB+13] gave an algorithm to recognize outer 1-planar graphs in linear time, is there some generalization or other strategy to extend this result to higher levels of non-planarity?

Open Problem 1. Can outer k-planarity for $k \geq 2$ be tested in polynomial time?

Outer 3-quasi-planar graphs are exactly the planar graphs, but we do not know of an efficient algorithm to test for outer k-quasi-planarity, even for $k = 3$.

Open Problem 2. Can outer k-quasi-planarity for $k > 3$ be tested in polynomial time?

Considering the coloring result for outer k-planar graphs, we would also be curious to know the answer to the following question:

Open Problem 3. Can the chromatic number of outer k-quasi-planar graphs be upper-bounded by some function k?

Polygonal Boundaries. In Chapter 4, we developed an algorithm to decide whether or not an outerplanar graph can be drawn into a given simple (not necessarily convex) polygon, when the vertices are mapped to the boundary and when each of the edges is allowed to have one bend inside the polygon. The algorithm we proposed works with the dual tree of the outerplanar graph, refining the polygon with each drawn edge until all edges are drawn or one of the bend points can not be placed inside the polygon.

A possible application for our algorithm is to recursively insert subgraphs into the inner faces of some predefined drawing. Being able to decide the question if a one-bend drawing for outerplanar graphs exists naturally gives rise to questions about related graph classes.

Open Problem 4. Can we decide if planar graphs can be drawn inside a simple polygon with one bend per edge when only some of the vertices are mapped to the polygon's boundary?

Open Problem 5. Can we incorporate crossings into the algorithm, deciding if outer k-planar graphs can be drawn with one bend per edge and inside a simple polygon?

Considering the bend points of our edges to be degree 2 subdivision vertices, we get a very restricted family of planar graphs drawable by our algorithm – finding a sufficient condition for larger subsets of planar graphs would be very interesting. Also, our algorithm works by iteratively refining the polygon – drawing an edge changes the boundary and planarity limits the options for further edges. What if we allowed some crossings on the inserted edges? Expanding the set of graph classes drawable by our algorithm – or similar strategies – would improve its applicability as a subroutine.

As our algorithm is stated now, it will add bend points to almost all edges of the output drawing to ensure that the refinements made to obtain the intermediate polygons are as little restrictive as possible. While this helped proving the algorithm's correctness, it can lead to drawings that are quite hard to read.

Open Problem 6. Can our algorithm be adapted to minimize the total number of bend points added to edges of the drawing?

Part Two: Drawing Vertices using Integer Coordinates

Moving to the Grid Optimally. Chapter 5 considered the task of transforming a given planar drawing into a topologically equivalent grid drawing with minimum vertex displacement – that is, a drawing with all vertices at integer coordinates, inducing the same embedding in the plane, and all vertices optimally cumulatively close to their original position. This task lends itself to trying rounding-like procedures such as modified variants of snap rounding, that are well-known in computational geometry.

We showed that TOPOLOGICALLY-SAFE GRID REPRESENTATION (as we call the problem above) is $\mathcal{NP}$-hard. Since we could not hope finding a modified version of an efficient snap rounding algorithm to solve it, we modelled TOPOLOGICALLY-SAFE GRID REPRESENTATION as an integer linear program. To evaluate the performance of our model, we implemented it in Java using the IBM CPLEX solver. We found our implementation to perform poorly on large instances – both area-wise and with respect to the number of vertices. While lazily generating the constraints for our model "on demand" gave some speedup, the model is still infeasible for any practical purpose. This raises the question about further improvements on the model.

Open Problem 7. As not all grid points will be used, can a column-generation-like strategy be used to iteratively "add" new grid points to the model?

Our analysis suggests that the area of the drawing has the largest impact on the wall-clock runtime of our implementation, because the sizes of most sets of constraints heavily depend on the number of possible edge slopes. One way of limiting the number of possible slopes would be to settle Open Problem 7.

As a byproduct, our implementation can also create minimum-area straight-line drawings of planar graphs – a problem also known to be $\mathcal{NP}$-hard [KW07]. Our experiments suggest that this can in practice only be used to verify or produce very small (counter-)examples. To the best of our knowledge, the problem of finding minimum-area straight-line drawings in reasonable time is still open.

Open Problem 8. Can we adapt the integer linear program solving TOPOLOGICALLY-SAFE GRID REPRESENTATION to find area-minimal straight-line planar drawings faster?

Rounding to the Grid Heuristically. In Chapter 6, we have presented a practical heuristic for the TOPOLOGICALLY-SAFE GRID REPRESENTATION problem, introduced in Chapter 5. Our algorithm follows the simulated annealing metaheuristic, but is subdivided into two distinct phases – one for feasibility, the other for optimization. The various features of the algorithm have been statistically evaluated, showing significant improvements to runtime or solution quality.

In Chapter 5, we presented a slow but exact algorithm, we gave a fast algorithm that is likely to produce reasonable results. The obvious open problems are trying to strengthen this rather vague statement: One could try finding an efficient deterministic heuristic or some approximation algorithm with provable guarantees.

Open Problem 9. Is there a deterministic heuristic or an approximation algorithm with provable guarantee for TOPOLOGICALLY-SAFE GRID REPRESENTATION?

While we obtained a result on approximation hardness (see Corollary 5.4 on page 70) for a special case, we failed to show $\mathcal{APX}$-hardness for the general TOPOLOGICALLY-SAFE GRID REPRESENTATION problem.

We have considered TOPOLOGICALLY-SAFE GRID REPRESENTATION as an abstract problem in isolation. Future work can consider the place of grid representations in a larger context. For example, we noted it may be useful to integrate polyline simplification for geographic data (recall the rather extreme examples in Figure 6.7 on page 109). It would also be interesting to evaluate the influence that rounding to a grid representation has on subsequent steps in a pipeline, such as the length of shortest paths, the results of generalization, or, for example, map matching.

Open Problem 10. Investigate how our two-stage heuristic algorithm can be adapted to allow for better perception of rounded geographic data.

Recognizing Nets of Orthogonal Polyhedra. In the final chapter of this book – Chapter 7 – we built upon the BUNDLECOLORING algorithm by Biedl and Genç [BG09]. Analyzing a repeated sequence of exhaustive executions of BUNDLECOLORING (which we called ITERATEDBUNDLECOLORING), we were able to translate Cauchy's Rigidity Theorem to orthogonal polyhedra of arbitrary genus and where the graph formed by the vertices and edges is connected – settling an open question proposed by Biedl and Genç. Given an unconnected graph, many distinct realizations are possible even for genus 0.[1] Our result encourages looking into other more diverse objects – like nets with facial angles that are multiples of 45°.

Open Problem 11. Is there a translation of our rigidity theorem to other polyhedra, for instance allowing dihedral and/or facial angles of that are multiples of 45°?

While our algorithm finds its output in a brute-force-like fashion, the analysis we used to show its correctness relied on carefully traversing a hypothetical realization, layer by layer. This resulted in a rather simple (and easily implementable) algorithm with a rather trivial upper bound of $O(m^3)$ on the runtime. This bound was obtained from the total number of possible restarts in each round and the maximum number of rounds, but our analysis uses only a rather small but well-selected subset of these executions.

Open Problem 12. Is there a better bound on the total number of restarts performed by ITERATEDBUNDLECOLORING?

It is also worth noting that neither our algorithm nor its runtime or its analysis are impacted by the actual genus of the resulting polyhedron. This is rather surprising, as the challenges seem to be related to the inner walls of the holes of the polyhedron. It seems possible that the number of restarts (and thus also the total runtime) can be parameterized by the genus of the realization. We have a family of instances that requires at least genus-many restarts: Recall that the object shown in Figure 7.2 (b) on page 117 requires two starts. Extending this construction by adding more rings onto the connecting bridge, we get objects of arbitrary genus that require equally many executions of BUNDLECOLORING. This raises the hope for a $O(g \cdot m^2)$ or even $O(g \cdot m)$-time algorithm.

---

[1] This fact was exploited by Biedl and Genç when they showed that the realization problem is $\mathcal{NP}$-hard for disconnected graphs.

Bibliography

[ABB+13] Christopher Auer, Christian Bachmaier, Franz J. Brandenburg, Andreas Gleißner, Kathrin Hanauer, Daniel Neuwirth, and Josef Reislhuber. Recognizing Outer 1-Planar Graphs in Linear Time. In Stephen K. Wismath and Alexander Wolff, editors, *Graph Drawing*, volume 8242 of *Lecture Notes in Computer Science*, pages 107–118. Springer International Publishing, 2013.

[ABB+16] Christopher Auer, Christian Bachmaier, Franz J. Brandenburg, Andreas Gleißner, Kathrin Hanauer, Daniel Neuwirth, and Josef Reislhuber. Outer 1-Planar Graphs. *Algorithmica*, 74(4):1293–1320, 2016.

[ABB+20] Patrizio Angelini, Michael A. Bekos, Franz J. Brandenburg, Giordano Da Lozzo, Giuseppe Di Battista, Walter Didimo, Michael Hoffmann, Giuseppe Liotta, Fabrizio Montecchiani, Ignaz Rutter, and Csaba D. Tóth. Simple k-planar graphs are simple $(k + 1)$-quasiplanar. *Journal of Combinatorial Theory, Series B*, 142:1–35, may 2020.

[ABS12] Evmorfia N. Argyriou, Michael A. Bekos, and Antonios Symvonis. The Straight-Line RAC Drawing Problem is NP-Hard. *Journal of Graph Algorithms and Applications*, 16(2):569–597, 2012.

[Ack09] Eyal Ackerman. On the Maximum Number of Edges in Topological Graphs with no Four Pairwise Crossing Edges. *Discrete and Computational Geometry*, 41(3):365–375, 2009.

[ADF+15] Patrizio Angelini, Giuseppe Di Battista, Fabrizio Frati, Vít Jelínek, Jan Kratochvíl, Maurizio Patrignani, and Ignaz Rutter. Testing Planarity of Partially Embedded Graphs. *ACM Transactions on Algorithms*, 11(4):1–42, 2015.

[AH76] K. Appel and W. Haken. Every planar map is four colorable. *Bulletin of the American Mathematical Society*, 82(5):711–713, 1976.

[AKL+20] Patrizio Angelini, Philipp Kindermann, Andre Löffler, Lena Schlipf, and Antonios Symvonis. One-Bend Drawings of Outerplanar Graphs Inside Simple Polygons. In Steven Chaplick, Philipp Kindermann, and Alexander Wolff, editors, *EuroCG2020*, 2020.

[AT07] Eyal Ackerman and Gábor Tardos. On the maximum number of edges in quasi-planar graphs. *Journal of Combinatorial Theory, Series A*, 114(3):563–571, 2007.

[AZ04]   Martin Aigner and Günter M. Ziegler. Cauchy's rigidity theorem. In *Proofs from THE BOOK*, pages 71–74. Springer Berlin Heidelberg, 2004.

[BBN+13]   Therese C. Biedl, Thomas Bläsius, Benjamin Niedermann, Martin Nöllenburg, Roman Prutkin, and Ignaz Rutter. Using ILP/SAT to Determine Pathwidth, Visibility Representations, and other Grid-Based Graph Drawings. In Stephen K. Wismath and Alexander Wolff, editors, *Graph Drawing*, volume 8242 of *Lecture Notes in Computer Science*, pages 460–471. Springer International Publishing, 2013.

[BCD+02]   Therese C. Biedl, Timothy M. Chan, Erik D. Demaine, Martin L. Demaine, Paul Nijjar, Ryuhei Uehara, and Ming-wei Wang. Tighter Bounds on the Genus of Nonorthogonal Polyhedra Built from Rectangles. In *14th CCCG*, pages 105–108, 2002.

[BE18]   Michael J. Bannister and David Eppstein. Crossing Minimization for 1-page and 2-page Drawings of Graphs with Bounded Treewidth. *Journal of Graph Algorithms and Applications*, 22(4):577–606, 2018.

[BEG+04]   Franz-Josef Brandenburg, David Eppstein, Michael T. Goodrich, Stephen G. Kobourov, Giuseppe Liotta, and Petra Mutzel. Selected Open Problems in Graph Drawing. In Giuseppe Liotta, editor, *Graph Drawing*, volume 2912 of *Lecture Notes in Computer Science*, pages 515–539. Springer Berlin Heidelberg, 2004.

[Ber83]   Jacques Bertin. *Semiology of graphics; diagrams networks maps*. University of Wisconsin Press, 1983.

[BG08]   Therese C. Biedl and Burkay Genç. Cauchy's Theorem for orthogonal polyhedra of genus 0. Technical Report CS-2008-26, University of Waterloo, School of Computer Science, 2008.

[BG09]   Therese C. Biedl and Burkay Genç. Cauchy's Theorem for Orthogonal Polyhedra of Genus 0. In Amos Fiat and Peter Sanders, editors, *European Symposium on Algorithms*, volume 5757 of *Lecture Notes in Computer Science*, pages 71–82. Springer Berlin Heidelberg, 2009.

[BG11]   Therese C. Biedl and Burkay Genç. Stoker's Theorem for Orthogonal Polyhedra. *International Journal of Computational Geometry & Applications*, 21(4):383–391, 2011.

[BGHL18]   Carla Binucci, Emilio Di Giacomo, Md. Iqbal Hossain, and Giuseppe Liotta. 1-page and 2-page drawings with bounded number of crossings per edge. *European Journal of Combinatorics*, 68(Supplement C):24–37, 2018.

[BH92]   Harry Buhrman and Steven Homer. Superpolynomial Circuits, Almost Sparse Oracles and the Exponential Hierarchy. In R. K. Shyamasundar, editor, *Lecture Notes in Computer Science*, volume 652 of *Lecture Notes in Computer Science*, pages 116–127. Springer Berlin Heidelberg, 1992.

[BKN16] Jasine Babu, Areej Khoury, and Ilan Newman. Every Property of Outerplanar Graphs is Testable. In Klaus Jansen, Claire Mathieu, José D. P. Rolim, and Chris Umans, editors, *Approximation, Randomization, and Combinatorial Optimization. Algorithms and Techniques (APPROX/RANDOM 2016)*, volume 60 of *Leibniz International Proceedings in Informatics (LIPIcs)*. Schloss Dagstuhl–Leibniz-Zentrum für Informatik, 2016.

[BLS05] Therese C. Biedl, Anna Lubiw, and Julie Sun. When can a net fold to a polyhedron? *Computational Geometry*, 31(3):207–218, 2005.

[BO79] Bentley and Ottmann. Algorithms for Reporting and Counting Geometric Intersections. *IEEE Transactions on Computers*, C-28(9):643–647, 1979.

[Bor84] O. V. Borodin. Solution of the Ringel problem on vertex-face coloring of planar graphs and coloring of 1-planar graphs. *Metody Diskret. Analiz.*, 41:12–26, 1984.

[Cab06] Sergio Cabello. Planar embeddability of the vertices of a graph using a fixed point set is NP-hard. *Journal of Graph Algorithms and Applications*, 10(2):353–363, 2006.

[CE12] Bruno Courcelle and J. Engelfriet. *Graph Structure and Monadic Second-Order Logic: A Language-Theoretic Approach*. Cambridge University Press, 2012.

[CFG+15] Timothy M. Chan, Fabrizio Frati, Carsten Gutwenger, Anna Lubiw, Petra Mutzel, and Marcus Schaefer. Drawing Partially Embedded and Simultaneously Planar Graphs. *Journal of Graph Algorithms and Applications*, 19(2):681–706, 2015.

[CFK+15] Marek Cygan, Fedor V. Fomin, Łukasz Kowalik, Daniel Lokshtanov, Dániel Marx, Marcin Pilipczuk, Michał Pilipczuk, and Saket Saurabh. *Parameterized Algorithms*, chapter Lower Bounds Based on the Exponential-Time Hypothesis, pages 467–521. Springer, 2015.

[CH67] Gary Chartrand and Frank Harary. Planar Permutation Graphs. *Annales de l'I.H.P. Probabilités et statistiques*, 3(4):433–438, 1967.

[Che93] L. Paul Chew. Guaranteed-quality mesh generation for curved surfaces. In *Proceedings of the ninth annual symposium on Computational geometry - SCG '93*. ACM Press, 1993.

[Chi08] John W. Chinneck. *Feasibility and Infeasibility in Optimization: Algorithms and Computational Methods*, volume 118 of *International Series in Operations Research & Management Science*. Springer US, 2008.

[CKL$^+$17] Steven Chaplick, Myroslav Kryven, Giuseppe Liotta, Andre Löffler, and Alexander Wolff. Beyond Outerplanarity. In Fabrizio Frati and Kwan-Liu Ma, editors, *Proceedings of the 25th International Symposium on Graph Drawing and Network Visualization.*, volume 10692 of *Lecture Notes in Computer Science*, pages 546–559. Springer, 2017.

[CLR87] Fan R. K. Chung, Frank Thomson Leighton, and Arnold L. Rosenberg. Embedding Graphs in Books: A Layout Problem with Applications to VLSI Design. *SIAM Journal on Algebraic and Discrete Methods*, 8(1):33–58, 1987.

[CLRS13] Thomas H. Cormen, Charles E. Leiserson, Ronald L. Rivest, and Clifford Stein. *Introduction to Algorithms*. MIT press, 3rd edition edition, 2013.

[CLWZ19] Steven Chaplick, Fabian Lipp, Alexander Wolff, and Johannes Zink. Compact drawings of 1-planar graphs with right-angle crossings and few bends. *Computational Geometry*, 84:50–68, 2019.

[CN98] Marek Chrobak and Shin-Ichi Nakano. Minimum-width grid drawings of plane graphs. *Computational Geometry*, 11(1):29–54, 1998.

[Con79] Robert Connelly. The Rigidity of Polyhedral Surfaces. *Mathematics Magazine*, 52(5):275–283, 1979.

[Coo71] Stephen A. Cook. The complexity of theorem-proving procedures. In *Proceedings of the third annual ACM symposium on Theory of computing - STOC '71*, pages 151–158. ACM Press, 1971.

[Cou90] Bruno Courcelle. The monadic second-order logic of graphs. I. Recognizable sets of finite graphs. *Information and Computation*, 85(1):12–75, 1990.

[CP92] Vasilis Capoyleas and János Pach. A Turán-type theorem on chords of a convex polygon. *Journal of Combinatorial Theory, Series B*, 56(1):9–15, 1992.

[CSW97] Robert Connelly, Idzhad Sabitov, and Anke Walz. The bellows conjecture. *Beiträge zur Algebra und Geometrie*, 38(1):1–10, 1997.

[CvDK$^+$20] Steven Chaplick, Thomas C. van Dijk, Myroslav Kryven, Ji won Park, Alexander Ravsky, and Alexander Wolff. Bundled Crossings Revisited. *Journal of Graph Algorithms and Applications*, 2020. (accepted, to be published).

[DBETT94] Giuseppe Di Battista, Peter Eades, Roberto Tamassia, and Ioannis G. Tollis. Algorithms For Drawing Graphs: An Annotated Survey. *Computational Geometry*, 4(5):235–282, 1994.

[dBHO07] Mark de Berg, Dan Halperin, and Mark Overmars. An intersection-sensitive algorithm for snap rounding. *Computational Geometry*, 36(3):159–165, 2007.

[dBK12] Mark de Berg and Amirali Khosravi. Optimal binary space partitions for segments in the plane. *International Journal of Computational Geometry & Applications*, 22(03):187–205, 2012.

[DEW17] Vida Dujmović, David Eppstein, and David R. Wood. Structure of Graphs with Locally Restricted Crossings. *SIAM Journal on Discrete Mathematics*, 31(2):805–824, 2017.

[dFPP90] Hubert de Fraysseix, János Pach, and Richard Pollack. How to draw a planar graph on a grid. *Combinatorica*, 10(1):41–51, 1990.

[DG02] Olivier Devillers and Pierre-Marie Gandoin. Rounding Voronoi diagram. *Theoretical Computer Science*, 283(1):203–221, 2002.

[Dil87] Michael B. Dillencourt. A non-Hamiltonian, nondegenerate Delaunay Triangulation. *Information Processing Letters*, 25(3):149–151, 1987.

[DKM02] Andreas W. M. Dress, Jack H. Koolen, and Vincent Moulton. On Line Arrangements in the Hyperbolic Plane. *European Journal of Combinatorics*, 23(5):549–557, 2002.

[DLL18] Olivier Devillers, Sylvain Lazard, and William J. Lenhart. 3D Snap Rounding. In Bettina Speckmann and Csaba D. Tóth, editors, *34th International Symposium on Computational Geometry (SoCG 2018)*, volume 99 of *Leibniz International Proceedings in Informatics (LIPIcs)*, pages 30:1–30:14, Dagstuhl, Germany, 2018. Schloss Dagstuhl–Leibniz-Zentrum fuer Informatik.

[DLM19] Walter Didimo, Giuseppe Liotta, and Fabrizio Montecchiani. A Survey on Graph Drawing Beyond Planarity. *ACM Computing Surveys*, 52(1):1–37, 2019.

[DLT83] Danny Dolev, Frank Thomson Leighton, and Howard Trickey. Planar Embedding of Planar Graphs. Technical report, Massachusetts Institute of Technology, 1983.

[DN19] Zdeněk Dvořák and Sergey Norin. Treewidth of graphs with balanced separations. *Journal of Combinatorial Theory, Series B*, 137:137–144, 2019.

[DO02]   Melody Donoso and Joseph O'Rourke. Nonorthogonal polyhedra built from rectangles. In *14th CCCG*, pages 101–104, 2002.

[DSW07]   Vida Dujmović, Matthew Suderman, and David R. Wood. Graph drawings with few slopes. *Computational Geometry*, 38(3):181–193, 2007.

[Ead84]   Peter Eades. A heuristic for graph drawing. *Congressus Numerantium*, 42:149–160, 1984.

[EM14]   David Eppstein and Elena Mumford. Steinitz Theorems for Simple Orthogonal Polyhedra. *Journal of Computational Geometry*, 5:179–244, 2014.

[EW90]   Peter Eades and Nicholas C. Wormald. Fixed edge-length graph drawing is NP-hard. *Discrete Applied Mathematics*, 28(2):111–134, 1990.

[Fár48]   István Fáry. On straight Lines representation of plane graphs. *ACTA Scientiarum Mathematicarum Szeged*, 11:229–233, 1948.

[FP07]   Fabrizio Frati and Maurizio Patrignani. A Note on Minimum-Area Straight-Line Drawings of Planar Graphs. In Seok-Hee Hong, Takao Nishizeki, and Wu Quan, editors, *Graph Drawing*, volume 4875 of *Lecture Notes in Computer Science*, pages 339–344. Springer Berlin Heidelberg, 2007.

[FPS13]   Jacob Fox, János Pach, and Andrew Suk. The Number of Edges in k-Quasi-planar Graphs. *SIAM Journal on Discrete Mathematics*, 27(1):550–561, 2013.

[FR91]   Thomas M. J. Fruchterman and Edward M. Reingold. Graph drawing by force-directed placement. *Software: Practice and Experience*, 21(11):1129–1164, 1991.

[Gas12]   William I. Gasarch. Guest Column "the second P =?NP poll". *ACM SIGACT News*, 43(2):53–77, 2012.

[GGHT97] Michael T. Goodrich, Leonidas J. Guibas, John Hershberger, and Paul J. Tanenbaum. Snap rounding line segments efficiently in two and three dimensions. In *Proceedings of the 13th Annual Symposium on Computational Geometry*, pages 284–293. ACM Press, 1997.

[Gil14]   Alexander Gilbers. *Visibility Domains and Complexity*. PhD thesis, Rheinische Friedrich-Wilhelms-Universität Bonn, 2014.

[GJ79]   Michael R. Garey and David S. Johnson. *Computers and Intractability: A Guide to the Theory of NP-Completeness*. W.H. Freeman, San Fransisco, CA, 2nd edition, 1979.

[GJ⁺10] Gaël Guennebaud, Benoît Jacob, et al. Eigen v3. http://eigen.tuxfamily.org, 2010.

[GKP94] Ronald L. Graham, Donald E. Knuth, and Oren Patashnik. *Concrete Mathematics: A Foundation for Computer Science*. Addison Wesley, 2nd edition, 1994.

[GKR94] V. Granville, M. Krivanek, and J.-P. Rasson. Simulated annealing: a proof of convergence. *IEEE Transactions on Pattern Analysis and Machine Intelligence*, 16(6):652–656, 1994.

[GKS92] Leonidas J. Guibas, Donald E. Knuth, and Micha Sharir. Randomized incremental construction of Delaunay and Voronoi diagrams. *Algorithmica*, 7(1–6):381–413, 1992.

[GKT14] Jesse Geneson, Tanya Khovanova, and Jonathan Tidor. Convex geometric $(k+2)$-quasiplanar representations of semi-bar k-visibility graphs. *Discrete Mathematics*, 331:83–88, 2014.

[GM98] Leonidas J. Guibas and David H. Marimont. Rounding Arrangements Dynamically. *International Journal of Computational Geometry & Applications*, 8(2):157–178, 1998.

[Gol91] David Goldberg. What every computer scientist should know about floating-point arithmetic. *ACM Computing Surveys*, 23(1):5–48, 1991.

[GY86] Daniel H. Greene and F. Frances Yao. Finite-resolution Computational Geometry. In *27th Annual Symposium on Foundations of Computer Science (SFCS 1986)*, pages 143–152. IEEE, 1986.

[HEK⁺14] Seok-Hee Hong, Peter Eades, Naoki Katoh, Giuseppe Liotta, Pascal Schweitzer, and Yusuke Suzuki. A Linear-Time Algorithm for Testing Outer-1-Planarity. *Algorithmica*, 72(4):1033–1054, 2014.

[Her13] John Hershberger. Stable snap rounding. *Computational Geometry*, 46(4):403–416, 2013.

[HN08] Seok-Hee Hong and Hiroshi Nagamochi. Convex drawings of graphs with non-convex boundary constraints. *Discrete Applied Mathematics*, 156(12):2368–2380, 2008.

[HN16] Seok-Hee Hong and Hiroshi Nagamochi. Testing Full Outer-2-planarity in Linear Time. In Ernst W. Mayr, editor, *Graph-Theoretic Concepts in Computer Science*, volume 9224, pages 406–421. Springer Berlin Heidelberg, 2016.

[Hob99] John D. Hobby. Practical segment intersection with finite precision output. *Computational Geometry*, 13(4):199–214, 1999.

[HP02] Dan Halperin and Eli Packer. Iterated snap rounding. *Computational Geometry*, 23(2):209–225, 2002.

[HS98] D. Harel and M. Sardas. An Algorithm for Straight-Line Drawing of Planar Graphs. *Algorithmica*, 20(2):119–135, 1998.

[IP01] Russell Impagliazzo and Ramamohan Paturi. On the Complexity of k-SAT. *Journal of Computer and System Sciences*, 62(2):367–375, 2001.

[JKR13] Vít Jelínek, Jan Kratochvíl, and Ignaz Rutter. A Kuratowski-type theorem for planarity of partially embedded graphs. *Computational Geometry*, 46(4):466–492, 2013.

[Kar72] Richard M. Karp. Reducibility among Combinatorial Problems. In *Complexity of Computer Computations*, pages 85–103. Springer US, 1972.

[KGV83] S. Kirkpatrick, C. D. Gelatt, and M. P. Vecchi. Optimization by Simulated Annealing. *Science*, 220(4598):671–680, 1983.

[KHN+14] Amruta Khot, Abdeltawab Hendawi, Anderson Nascimento, Raj Katti, Ankur Teredesai, and Mohamed Ali. Road network compression techniques in spatiotemporal embedded systems. In *5th ACM SIGSPATIAL International Workshop on GeoStreaming - (IWGS '14)*. ACM Press, 2014.

[KKRW10] Bastian Katz, Marcus Krug, Ignaz Rutter, and Alexander Wolff. Manhattan-Geodesic Embedding of Planar Graphs. In David Eppstein and Emden R. Gansner, editors, *Graph Drawing*, volume 5849 of *Lecture Notes in Computer Science*, pages 207–218. Springer Berlin Heidelberg, 2010.

[KLM17] Stephen G. Kobourov, Giuseppe Liotta, and Fabrizio Montecchiani. An annotated bibliography on 1-planarity. *Computer Science Review*, 25:49–67, 2017.

[Kra11] Karl Kraus. *Photogrammetry: Geometry from Images and Laser Scans*. Walter de Gruyter, 2011.

[KW07] Marcus Krug and Dorothea Wagner. Minimizing the Area for Planar Straight-Line Grid Drawings. In Seok-Hee Hong, Takao Nishizeki, and Wu Quan, editors, *Graph Drawing*, volume 4875 of *Lecture Notes in Computer Science*, pages 207–212. Springer Berlin Heidelberg, 2007.

[Lev73] Leonid A. Levin. Universal sequential search problems. *Problemy Peredachi Informatsii*, 9(3):115–116, 1973.

[LMM18] Anna Lubiw, Tillmann Miltzow, and Debajyoti Mondal. The Complexity of Drawing a Graph in a Polygonal Region. In Therese C. Biedl and Andreas Kerren, editors, *Proc. 26th Int. Symp. Graph Drawing Netw. Vis.*, volume 11282, pages 387–401. Springer International Publishing, 2018.

[Löf16]   Andre Löffler.   Snapping Graph Drawings to the Grid.   Master's thesis, Julius-Maximilians-Universität Würzburg, 2016.   Available at `http://www1.pub.informatik.uni-wuerzburg.de/pub/theses/2017-loeffler-master.pdf`.

[LvDW16] Andre Löffler, Thomas van Dijk, and Alexander Wolff. Snapping Graph Drawings to the Grid Optimally. In *Proceedings of the 24th International Symposium on Graph Drawing*, volume 9801 of *Lecture Notes in Computer Science*, pages 144–151. Springer International Publishing, 2016.

[LW70]   Don R. Lick and Arthur T. White. k-Degenerate Graphs. *Canadian Journal of Mathematics*, 22:1082–1096, 1970.

[Mil95]   Victor J. Milenkovic.   Practical methods for set operations on polygons using exact arithmetic. In *7th CCCG*, pages 55–60, 1995.

[MKNF87] Sumio Masuda, Toshinobu Kashiwabara, Kazuo Nakajima, and Toshio Fujisawa.   On the NP-completeness of a computer network layout problem. In *Proceedings of the 1987 IEEE International Symp. on Circuits and Systems*, pages 292–295, 1987.

[MN90]   Victor J. Milenkovic and Lee R. Nackman.   Finding compact coordinate representations for polygons and polyhedra. *IBM Journal of Research and Development*, 34(5):753–769, 1990.

[MNR13]   Tamara Mchedlidze, Martin Nöllenburg, and Ignaz Rutter. Drawing Planar Graphs with a Prescribed Inner Face. In Stephen K. Wismath and Alexander Wolff, editors, *Graph Drawing*, volume 8242 of *Lecture Notes in Computer Science*, pages 316–327. Springer International Publishing, 2013.

[MS97]   Bruce A. McCarl and Thomas H. Spreen. *Applied Mathematical Programming Using Algebraic Systems*. Texas A&M University, 1997.

[MU18]   Tamara Mchedlidze and Jérôme Urhausen. β-Stars or On Extending a Drawing of a Connected Subgraph. In Therese C. Biedl and Andreas Kerren, editors, *Proc. 26th Int. Symp. Graph Drawing Netw. Vis.*, volume 11282, pages 416–429, 2018.

[Nak00]   Tomoki Nakamigawa. A generalization of diagonal flips in a convex polygon. *Theoretical Computer Science*, 235(2):271–282, 2000.

[NDG$^+$16] Alberto Noronha, Anna Dröfn Daníelsdóttir, Piotr Gawron, Freyr Jóhannsson, Soffía Jónsdóttir, Sindri Jarlsson, Jón Pétur Gunnarsson, Sigurður Brynjólfsson, Reinhard Schneider, Ines Thiele, and Ronan M. T. Fleming. ReconMap: an interactive visualization of human metabolism. *Bioinformatics*, 33(4):605–607, 2016.

[Nöl05] Martin Nöllenburg. Automated Drawing of Metro Maps. Master's thesis, Fakultät für Informatik, Universität Karlsruhe, 2005. Available at `https://i11www.iti.kit.edu/extra/publications/n-admm-05da.pdf`.

[NW11] Martin Nöllenburg and Alexander Wolff. Drawing and Labeling High-Quality Metro Maps by Mixed-Integer Programming. *IEEE Transactions on Visualization and Computer Graphics*, 17(5):626–641, 2011.

[Pac06] Eli Packer. Iterated snap rounding with bounded drift. In *Proceedings of the 22nd Annual Symposium on Computational Geometry*, pages 367–376. ACM Press, 2006.

[Pac19] Eli Packer. 2D Snap Rounding. In *CGAL User and Reference Manual*. CGAL Editorial Board, 4.14 edition, 2019.

[Pat06] Maurizio Patrignani. On Extending a Partial Straight-line Drawing. *International Journal of Foundations of Computer Science*, 17(5):1061–1069, 2006.

[PS85] Franco P. Preparata and Michael Ian Shamos. *Computational Geometry*. Springer New York, 1985.

[PS86] Andrzej Proskurowski and Maciej Syslo. Efficient Vertex- and Edge-Coloring of Outerplanar Graphs. *SIAM Journal on Algebraic and Discrete Methods*, 7:131–136, 1986.

[PSS96] J. Pach, F. Shahrokhi, and M. Szegedy. Applications of the crossing number. *Algorithmica*, 16(1):111–117, 1996.

[PT97] János Pach and Géza Tóth. Graphs drawn with few crossings per edge. *Combinatorica*, 17(3):427–439, 1997.

[PW01] János Pach and Rephael Wenger. Embedding Planar Graphs at Fixed Vertex Locations. *Graphs and Combinatorics*, 17(4):717–728, 2001.

[PW14] Dongliang Peng and Alexander Wolff. Watch Your Data Structures! In *Proceedings of the 22nd Annual Conference of the GIS Research UK*, pages 371–381, 2014.

[Rin65] Gerhard Ringel. Ein Sechsfarbenproblem auf der Kugel. *Abhandlungen aus dem Mathematischen Seminar der Universität Hamburg*, 29(1):107–117, 1965.

[RS84] Neil Robertson and Paul D. Seymour. Graph minors. III. Planar tree-width. *Journal of Combinatorial Theory, Series B*, 36(1):49–64, 1984.

[Sch90] Walter Schnyder. Embedding Planar Graphs on the Grid. In David S. Johnson, editor, *Proceedings of the first annual ACM-SIAM symposium on Discrete algorithms.*, pages 138–148. Society for Industrial and Applied Mathematics, 1990.

[Sch13] Marcus Schaefer. The Graph Crossing Number and its Variants: A Survey. *Electronic Journal of Combinatorics*, 1000, 2013.

[SE05] Niklas Sorensson and Niklas Een. MiniSat v1. 13 – A Sat Solver with Conflict-Clause Minimization. *Theory and Applications of Satisfiability Testing*, (53):1–2, 2005.

[SHDZ02] Shashi Shekhar, Yan Huang, Judy Djugash, and Changqing Zhou. Vector map compression. In *Proceedings of the tenth ACM International Symposium on Advances in Geographic Information Systems*. ACM Press, 2002.

[STT81] Kozo Sugiyama, Shojiro Tagawa, and Mitsuhiko Toda. Methods for Visual Understanding of Hierarchical System Structures. *IEEE Transactions on Systems, Man, and Cybernetics*, 11(2):109–125, 1981.

[TSF+13] Ines Thiele, Neil Swainston, Ronan M. T. Fleming, Andreas Hoppe, Swagatika Sahoo, Maike K. Aurich, Hulda Haraldsdottir, Monica L. Mo, Ottar Rolfsson, Miranda D. Stobbe, Stefan G. Thorleifsson, Rasmus Agren, Christian Bölling, Sergio Bordel, Arvind K. Chavali, Paul Dobson, Warwick B. Dunn, Lukas Endler, David Hala, Michael Hucka, Duncan Hull, Daniel Jameson, Neema Jamshidi, Jon J. Jonsson, Nick Juty, Sarah Keating, Intawat Nookaew, Nicolas Le Novère, Naglis Malys, Alexander Mazein, Jason A. Papin, Nathan D. Price, Evgeni Selkov, Martin I. Sigurdsson, Evangelos Simeonidis, Nikolaus Sonnenschein, Kieran Smallbone, Anatoly Sorokin, Johannes H. G. M. van Beek, Dieter Weichart, Igor Goryanin, Jens Nielsen, Hans V. Westerhoff, Douglas B. Kell, Pedro Mendes, and Bernhard Ø. Palsson. A community-driven global reconstruction of human metabolism. *Nature Biotechnology*, 31(5):419–425, 2013.

[Tut63] William T. Tutte. How to Draw a Graph. *Proceedings of the London Mathematical Society*, 13(1):743–767, 1963.

[vDH14] Thomas C. van Dijk and Jan-Henrik Haunert. Interactive focus maps using least-squares optimization. *International Journal of Geographical Information Science*, 28(10):2052–2075, 2014.

[vDL18] Thomas C. van Dijk and Dieter Lutz. Realtime linear cartograms and metro maps. In *Proceedings of the 26th ACM SIGSPATIAL International Conference on Advances in Geographic Information Systems*, pages 488–491. ACM Press, 2018.

[vDL19] Thomas C. van Dijk and Andre Löffler. Practical Topologically Safe Rounding of Geographic Networks. In *Proceedings of the 27th ACM SIGSPATIAL International Conference on Advances in Geographic Information Systems*, pages 239–248. ACM Press, 2019.

[vDvGH+13] Thomas van Dijk, Arthur van Goethem, Jan-Henrik Haunert, Wouter Meulemans, and Bettina Speckmann. Accentuating focus maps via partial schematization. In *Proceedings of the 21st ACM SIGSPATIAL International Conference on Advances in Geographic Information Systems*, pages 428–431. ACM Press, 2013.

[vLA87] Peter J. M. van Laarhoven and Emile H. L. Aarts. *Simulated Annealing: Theory and Applications*. Springer Netherlands, 1987.

[Wil45] Frank Wilcoxon. Individual Comparisons by Ranking Methods. *Biometrics Bulletin*, 1(6):80, 1945.

[WT07] David R. Wood and Jan Arne Telle. Planar decompositions and the crossing number of graphs with an excluded minor. *New York Journal of Mathematics*, 13:117–146, 2007.

[Yvi19] Mariette Yvinec. 2D Triangulation. In *CGAL User and Reference Manual*. CGAL Editorial Board, 4.14 edition, 2019. `https://doc.cgal.org/4.14/Manual/packages.html#PkgTriangulation2`.

[Zie08] Günter M. Ziegler. *Polyhedral Surfaces of High Genus*, pages 191–213. Birkhäuser Basel, Basel, 2008.

[ZWF19] Baruch Zukerman, Ron Wein, and Efi Fogel. 2D Intersection of Curves. In *CGAL User and Reference Manual*. CGAL Editorial Board, 4.14 edition, 2019. `https://doc.cgal.org/4.14/Manual/packages.html#PkgSurfaceSweep2`.

Acknowledgments

Somebody once told me that doing research is like shooting an arrow into the sky, then looking for where it came down to paint a bullseye around it, declaring victory. In retrospect, i fully agree.

Following this metaphor, first and foremost, I want to thank my supervisor Dr. Steven Chaplick for helping with painting the bullseye. Graph drawing and computational geometry are fields with problems that can be tackled from many different angles. Guiding me, he always managed to keep me focused on the task at hand. Knowledgeable and always happy to help, he really was the supervisor I needed to conclude this work.

All this work would not have been possible had Prof. Dr. Alexander Wolff not provided the arrow. He gave me a place in his group and took care of financing my four-year long archery lesson. But most importantly, his critical thinking and helpful comments as my senior advisor pushed me forward, making this shot possible.

Having an arrow would be rather pointless without also having a bow. Dr. Thomas C. van Dijk sparked my joy for research-archery, encouraging me to pick up my own bow. Leading by example, his curiosity made me to broaden my horizon, helping me search for that arrow after it came down in unexpected places.

For helping me in pulling the bowstring and watching the arrow fly, I thank all my coauthors (in alphabetical order): Patrizio Angelini, Moritz Beck, Johannes Blum, Steven Chaplick, Thomas C. van Dijk, Tobias Greiner, Bas den Heijer, Nadja Henning, Philipp Kindermann, Felix Klesen, Myroslav Kryven, Giuseppe Liotta, Lena Schlipf, Antonios Symvonis, Florian Thiele, Alexander Wolff, Alexander Zaft, and Johannes Zink. Working with all of you has been a pleasure, your ideas and contributions have been truly invaluable.

Research hardly ever happens in isolation. The wonderful people in the group of Würzburg had all the right impact on me, even if we didn't publish together. Thank you to Benedikt Budig, Titus Dose, Oksana Firman, Krzysztof Fleszar, Jakob Geiger, Christian Glaßer, Jonathan Klawitter, Fabian Lipp, Falco Nogatz, Ludwig Ostermayer, Dongliang Peng, Dietmar Seipel, Joachim Spoerhase, Sabine Storandt, and Daniel Weidner.

Finally, this undertaking would not have been possible without the wonderful support by my wife Eva and our son Felix. Without the two of you, I would have lost track and gone crazy a long time ago.

List of Publications

- A. Löffler, T. C. van Dijk, A. Wolff.
  Snapping Graph Drawings to the Grid Optimally.
  In: *Proceedings of the 24th International Symposium on Graph Drawing and Network Visualization 2016 (GD'16).* pp. 144-151. Springer (2016).

- S. Chaplick, M. Kryven, G. Liotta, A. Löffler, A. Wolff.
  Beyond Outerplanarity.
  In: *Proceedings of the 25th International Symposium on Graph Drawing and Network Visualization 2017 (GD'17).* pp. 546-559. Springer (2017).

- T. C. van Dijk, T. Greiner, B. den Heijer, N. Henning, F. Klesen, A. Löffler.
  Wüpstream: Efficient Enumeration of Upstream Features (GIS cup).
  In: *Proceedings of the 26th ACM SIGSPATIAL International Conference on Advances in Geographic Information Systems (SIGSPATIAL'18).* pp. 626-629. ACM (2018).

- M. Beck, J. Blum, M. Kryven, A. Löffler, J. Zink.
  Planar Steiner Orientation is NP-complete.
  At: *10th International Colloquium on Graph Theory and Combinatorics (ICGT'18).* Lyon, France, 2018.

- S. Chaplick, P. Kindermann, A. Löffler, F. Thiele, A. Wolff, A. Zaft, J. Zink.
  Stick Graphs with Length Constraints.
  In: *Proceedings of the 27th International Symposium on Graph Drawing and Network Visualization (GD'19).* pp. 3-17. Springer (2019).

- T. C. van Dijk, A. Löffler.
  Practical Topologically Safe Rounding of Geographic Networks.
  In: *Proceedings of the 27th ACM SIGSPATIAL International Conference on Advances in Geographic Information Systems (SIGSPATIAL'19).* pp. 239-248. ACM (2019).

- P. Angelini, P. Kindermann, A. Löffler, L. Schlipf, A. Symvonis.
  One-Bend Drawings of Outerplanar Graphs Inside Simple Polygons.
  In: *Proceedings of the 36th European Workshop on Computational Geometry (EuroCG'20).* pp 70:1–70:6, Würzburg (2020).